THE COMMON AGRICULTURAL POLICY AND ROMANIAN AGRICULTURE

THE COMMON AGRICULTURAL POLICY AND ROMANIAN AGRICULTURE

Jean Vasile Andrei

Business Administration Department, Petroleum-Gas University of Ploiești, 39 București Blvd, Ploiești, 100680, Prahova, Romania

and

Mihaela Cristina Drăgoi

Department of International Business and Economics, Bucharest University of Economic Studies, 6 Piața Romană, 1st District, Bucharest, 010374 Romania

CABI is a trading name of CAB International

CABI	CABI
Nosworthy Way	745 Atlantic Avenue
Wallingford	8th Floor
Oxfordshire OX10 8DE	Boston, MA 02111
UK	USA
Tel: +44 (0)1491 832111	Tel: +1 (617)682-9015
Fax: +44 (0)1491 833508	E-mail: cabi-nao@cabi.org
E-mail: info@cabi.org	
Website: www.cabi.org	

A catalogue record for this book is available from the British Library, London, UK.

Library of Congress Cataloging-in-Publication Data

Names: Jean-Vasile, Andrei, 1982- author. | Draăgoi, Mihaela Cristina, author.
Title: The Common Agricultural Policy and Romanian agriculture / Jean Vasile Andrei,
    Mihaela Cristina Dragoi.
Description: Boston, MA : CAB International, 2019. | Includes bibliographical references and
    index. | Summary: "The current evolution of the European economy suggests that in the near
    future, research in agri-food economy and agri-food production, and agricultural production
    systems and structures must be redesigned, adapted and developed to respond to the lack of
    sustainability of agri-food production systems and the current global food crisis. This book
    analyses the agricultural paradigm transformations that occur as countries converge on
    the European agricultural model and what their impact is for sectoral development, while
    emphasizing their contribution to the redefinition of rural agricultural communities and
    economy. This book helps develop a theoretical framework by analysing the specialized
    empirical literature and techniques used in the field of agricultural economy research, with
    a focus on the transformation of Romanian agriculture in order to become integrated and
    respond to the globalization of markets. presents, analyses and discusses the main theories in
    field of agricultural economics and paradigms; creates a working paradigm for this concept
    within agricultural economics; provides a theoretical framework for the agricultural model.
    The book is aimed at students and researchers in agricultural economics, and government and
    policy makers internationally"-- Provided by publisher.
Identifiers: LCCN 2019028807 (print) | LCCN 2019028808 (ebook) | ISBN 9781789242201
    (hardcover) | ISBN 9781789242218 (ebook) | ISBN 9781789242225 (epub)
Subjects: LCSH: Common Agricultural Policy. | Agriculture and state--Economic aspects--Europe. |
    Agriculture and state--Economic aspects--Romania--Case studies. | Agriculture--Economic
    aspects--Research--Europe. | Agriculture--Economic aspects--Research--Romania--Case studies.
Classification: LCC HD1918 .A53 2019 (print) | LCC HD1918 (ebook) | DDC 338.1/8498--dc23
LC record available at https://lccn.loc.gov/2019028807
LC ebook record available at https://lccn.loc.gov/2019028808

ISBN-13: 978 1 78924 220 1 (hardback)
        978 1 78924 221 8 (ePDF)
        978 1 78924 222 5 (ePub)

Commissioning Editor: David Hemming
Editorial Assistant: Emma McCann
Production Editor: Kate Hill

Typeset by SPi, Pondicherry, India
Printed and bound in the UK by Severn, Gloucester

Contents

About the Authors

Jean Vasile Andrei is an Associate Professor at Petroleum-Gas University of Ploieşti, Department of Business Administration, and a PhD mentor in economics at Bucharest University of Economic Studies, Romania. He holds a PhD in Economics from the National Institute of Economics Research – Romanian Academy of Sciences. He has earned a BA degree in Administrative Sciences (2005) and in Banks and Finances (2007) from the Petroleum-Gas University of Ploieşti. He has an MA degree in Economics, Administrative and Business Management (2007) earned at the same university. He is Editor-in-Chief of the *International Journal of Sustainable Economies Management* (USA), Associate Editor of *Economics of Agriculture* (*Ekonomikapoljoprivrede*) published by the Institute of Agricultural Economics (Serbia) and a scientific reviewer for the *International Business Information Management Association* Conferences (IBIMA). He is also a member of the Balkan Scientific Association of Agrarian Economists, Serbia (December 2008). Issues such as agricultural and resources economics, business administration, and valuing economic and human potential are among his research and scientific interests, having published over 50 articles and nine scientific books, and given numerous conference presentations.

Mihaela Cristina Drăgoi is an Associate Professor at the Faculty of International Business and Economics within the Bucharest University of Economic Studies in Romania. She studied at the Universitat Autonoma de Barcelona (Spain) for the last academic year of the bachelor's programme and then graduated from the Faculty of International Business and Economics with a specialization in International Business. She holds a master's degree in Marketing and Business Communication, and she obtained her PhD in the field of International Economic Relations from the Bucharest University of Economic Studies with a thesis entitled 'Optimization of the Romanian health care system in the context of European integration'. She continued her research within postdoctoral programmes in Bucharest (Romania), New York (USA) and Rome (Italy), and has published numerous works in her fields of interest regarding European integration, economic development, health economics and international business.

List of Figures

List of Tables

1 Introduction

The European economy is undergoing a profound process of adaptation to the new realities and constraints imposed by transformations in the overall global economy, and implicitly by the expanding phenomena of globalization and the deep integration of markets. Against this background, European agriculture faces a series of new challenges of increased amplitude, which must be overcome through profound changes in existing economic and social paradigms, by identifying optimal ways out of the crisis, as well as by regaining competitive advantages and the confirmation of a leading role in a market that is constantly being transformed. The development of the national agricultural sector under the auspices and rigours imposed by the Common Agricultural Policy (CAP) has led to a sustained increase in the level of sectoral competitiveness but also to numerous economic advantages for the rural population.

Recent developments in the agricultural sector and the agrarian economy generally point to the existence of potential imbalances, with strong anchors in the perturbations caused not only by the manifestation of a surplus of agricultural production and an increased pressure on price evolution but also by the accentuation of politically motivated interdictions or limitations and financial incentives.

The social effects generated by the influence of the agricultural sector on the creation of added value and the significant share of the agricultural rural population are complex and require a multidimensional approach. Hence, the role of agriculture in the economy as a whole must be analysed not only from the perspective of exploiting the existing agricultural and rural potential, or from the perspective of agricultural producers, but including the entire system of relations and determinations generated by this economic branch.

Agriculture is increasingly showing its multifunctional character, emphasizing its role and importance in the rural economy. The important transformations that marked the evolution of the agri-food and agricultural sector, amid diversification of production relations and the globalization and integration of markets, as well as the requirements of ensuring convergence with the European agricultural model and compatibility with the functioning mechanism of the CAP, have required a reconsideration of the role and place of this sector within the national economy as a whole.

The evolution of European agriculture is closely linked to the evolution of the CAP, which, while retaining to a certain extent its initial objectives, along with lengthy reforms, has had to adapt to the new realities and to develop a mechanism in order to capitalize on the economic and social potential of rural areas and agriculture. From this perspective, the CAP has established and developed over time a valid European agri-food model inseparable from the evolution of this sectoral policy. Against this background, the European agri-food model has thus had to go through and adapt to all the requirements imposed by CAP reforms, from the reform of instruments and management systems, funding and budgets, and rural development to opening up to third countries and new market mechanisms.

The CAP is therefore designed to support the development of a genuine and original European agriculture model, determined by the existence of a significant number of family farms with a substantial influence and traditional features, and of cooperatives and agricultural societies, which show a willingness to cooperate in agriculture. At the same time, the CAP needs to undertake a functional distribution of sectoral effects and to mobilize existing rural community resources, helping to reduce the wide imbalances and inequalities that occur between rural and urban areas, in terms of both income and socioeconomic importance.

Based on its structure of two pillars, the CAP addresses certain dimensions with a fundamental impact, from both the economic and the sectoral perspective, through the mechanisms developed in Pillar I, as well as the development of rural communities and the promotion of rural development, encouraged through the specific measures of Pillar II. The philosophy of building and operating the CAP must pivot around the two traditional principles, namely the principle of community preference and the principle regarding balanced food sovereignty from a territorial perspective, which guarantee both a sustainable supply of food to citizens and functionality at the optimal level of European agriculture.

Direct payments made under the CAP not only provide financial provisions and farm income support but inherently stimulate the development and re-engineering of agriculture, thus emphasizing the sectoral importance of this particular policy. Considering the cumulative effects of the agricultural policy measures determined by the rural development payments under the second pillar, the CAP contributes directly to supporting vulnerable regions and sectors and to reducing regional inequalities by promoting economic, social and environmental programmes, with an impact on rural communities.

Addressing a subject like the paradigm changes of agriculture in general, and of Romanian agriculture in particular, in the context of convergence with the European agricultural model is motivated by the fact that agriculture is still a determinant economic sector in contemporary economies with multiple influences and reverberations at various levels.

The agricultural sector is undergoing a broad and continuous process of linking to the changes in the European agricultural model, both at the level of CAP philosophy and paradigm modifications, and at the level of rural communities. Agriculture, although at the confluence of traditional and modern methods in contemporary economies, performs multiple functions with trans-sectoral effects.

To obtain comparable and long-lasting results, analysis of the paradigm shift in Romanian agriculture needs to be done in a proper, balanced manner, taking into account the possible transformations of the sectoral paradigm. Thus, it becomes important in this context to understand how agriculture, agricultural research as a whole, agri-food production, the structures of agricultural production and sustainability of rural communities can be improved at different levels in contemporary economies. In the field of agri-food production systems, agricultural production structures, food economy and sustainable rural development, an integrated approach is required, providing solutions for increasing the capitalization of the national agricultural potential but also for improving sectoral competitiveness.

The main objective of this book is an in-depth analysis of the agricultural paradigm transformations in the context of convergence with the European agricultural model, and the influences they impose on the process of sectoral development, while emphasizing their contribution to the redefinition of rural agricultural communities and economy.

Therefore, this book presents a synthesis of the evolution of the European agricultural policy and the transformation of the European agricultural sector in the context of modification of the paradigm of the agricultural model. At the same time, it addresses three fundamental themes, closely connected to the contemporary exigencies and challenges of the agri-food economy and rural space, as well as to the current research trends in the field. The current evolution of the European economy indicates that, in the near future, research in the field of agri-food economy and agri-food production, agricultural production systems and structures must be designed, adapted and developed to respond to the great challenge that the contemporary world is facing – the lack of sustainability of agri-food production systems and the food crisis.

This book aims to contribute to development of the theoretical framework specific to the field of research by presenting and critically analysing the specialized empirical literature as well as recent methods, tools and achievements in the field of agricultural economy research, and the systems and structures of agricultural production, as well as highlighting the complementary connections that this field develops.

In the sections dedicated to the dimensions, themes and approached research directions, the main transformations of the Romanian agriculture are highlighted in terms of integration and globalization of the markets, the influence of paradigm changes in the use of land, and the competitiveness of the national agricultural products in the community space, as well as the convergence of the Romanian agricultural model with the European model.

This book also attempts to present from a multidimensional perspective the role of the CAP and sectoral agricultural policies on the development of national agriculture, bringing to the fore the need to refine specific sectoral policies and adapt them to the realities required by the modification of funda-mental economic paradigms.

Highlighting the impact of the CAP on the evolution of Romanian agri-culture and its convergence with the European agricultural model is a topic of great relevance in the context of transformations and reforms of the European agricultural sectoral policy. The need to carry out such an analysis is also imposed by the need to raise awareness regarding the role of the CAP in shap-ing the national agricultural sector and complying with the requirements of a functional market economy in Romania.

Starting from the above-described arguments, this book aims to join and complete the field studies through a general-to-specific approach, reviewing both the evolution and the transformations of the CAP throughout its history, while providing in the final part an econometric study on the possibility of convergence of a national agricultural sector, with a significant influence in ensuring sustainable economic growth, such as that of Romania.

2 The Common Agricultural Policy: Long-standing and New Paradigms

The European Communities were created during the 1950s, a period that was defined, among various characteristics, by a shortage of agricultural products. World War II had left Europe impoverished and dependent on imports of agricultural products, mainly from the USA.

Upon creating the European Economic Community, the six Member States at that time[1] realized the importance of developing a common market for producing and selling agricultural products, thus diminishing the dependence on imports from outside the Community and creating an autonomous agricultural sector within the Community. The main goals that were pursued by the artisans of the common agricultural market were to ensure a self-sufficient production of agricultural goods, while also providing constant well-being for farmers. Thus, the bases of the CAP were established after the European Economic Community was formed. Additionally, when the future common market was established by the Treaty of Rome, the agriculture of the six founding Member States was characterized by a strong state intervention. In order to include agricultural products in the free movement of goods while maintaining state intervention in the agricultural sector, it was necessary to eliminate the national intervention mechanisms incompatible with the common market and to transpose them at Community level: this was the main reason for the occurrence of the CAP (European Parliament, 2017).

However, the CAP served a broader set of objectives, as stated by the Treaty on the Functioning of the European Union[2] – article 39 (Official Journal of the European Union, 2012):

1. Increasing agricultural productivity by promoting technical progress, by ensuring the rational development of agricultural production, and by optimizing the use of production factors, particularly of labour.
2. Ensuring a fair standard of living for the population involved in agriculture, in particular by increasing the individual income of agricultural workers.

3. Stabilization of the markets.
4. Ensuring security of supply.
5. Ensuring reasonable delivery prices for consumers.

Therefore, the purposes of the CAP are both economic and social, aiming to equally protect the interests of producers and consumers. In practice, the objectives of the CAP have remained unchanged since the Treaty of Rome, but the reformation of this policy, started in the 1980s, brought to our attention other important aspects that the CAP must deal with, apart from simply providing a common agricultural market for farmers and consumers. Thus, the reformed CAP also focuses on environmental protection, promoting a high level of employment, sustainable development, livestock well-being requirements and the protection of public health.

Although agriculture represents only a small fraction of the economies of developed countries today, including the European Union (EU), public intervention has recently been strengthened through agri-rural policies that have supplemented the traditional function of primary activity – namely the production of food for consumption – with other dimensions, including sustainable development, climate change mitigation, land and landscape planning, diversification and revitalization of the rural economy, and energy and biomaterial production. Supporting public goods, known as non-commercial functions of agricultural activity (i.e. unpaid), has therefore become a key element of more recent agricultural and rural policies, including the CAP (European Parliament, 2017).

2.1   Evolution of the CAP Over Time

Having had a pioneering role in the integration process, the CAP can be said to be one of the foundations of the EU today (European Institute of Romania, 2003). According to the principles set out in 1957 in the Treaty of Rome and listed above, the CAP measures and instruments were designed by the European Commission, through Sicco Mansholt, European Commissioner for Agriculture and Vice-President of the European Commission at the time (European Commission, 2017d). Basically, the mechanism consists of protecting farmers' incomes by means of prices, namely by establishing a high level of customs protection against foreign competition combined with the unification of domestic prices, specifically fixing common prices accompanied by a mechanism to support their levels. As domestic prices were higher than world prices, exports had to be encouraged through subsidies. These common measures would be financed from a common budget. The Community's agricultural market thus became a solid construction, with developments independent of the trends of international markets (European Institute of Romania, 2003).

Specifically, the CAP was based initially on the common market organizations, which comprised all the market instruments for the functioning of the single agricultural market, namely fixed prices, market intervention, common customs protection, production quotas and financial aid for farmers, as laid down by the Conference held in Stresa (Italy) in 1958. It was then established that the family farm must remain at the heart of European agriculture, which continues to mentally shape the structure of ample debate (Luca, 2009).

The general mechanism for the functioning of the agricultural markets in the EU was based on a complex system of regulating the prices of agricultural products. Thus, the following price levels for products falling under the CAP were set annually by the Council: the indicative price and the intervention price (Zăpodeanu and Popovici, 2006).

The indicative price, set higher than the international prices of agricultural products, was the highest level recommended by the Council for selling agricultural products within the Community, as it was considered to ensure a 'reasonable' standard of living for farmers; the intervention price was the guaranteed minimum price that could be obtained for domestically marketed production (Zăpodeanu and Popovici, 2006). When the price of a certain agricultural product reached the minimum level, explicitly the intervention price, signifying a lower demand compared with the offer, the Community intervened by purchasing and storing the product, thus re-establishing the market equilibrium. Purchases and storage were supported by the European Agricultural Guidance and Guarantee Fund (EAGGF). In this case, exports were also encouraged, but as the international prices of agricultural products were lower than the levels set for the intervention price and the indicative price, the EAGGF also supported the export subsidies for farmers, namely the difference between the international price and the intervention price of the agricultural products exported.

Imports for certain products were allowed in contingent form only when the price of agricultural products tended to rise above the indicative price, which signified that the Community offer was not sufficient to cover the demand for agricultural products. In this case, imports were subject to variable import levies, which raised the prices beyond the Community prices, in order to ensure the consumers' preference for local products instead of imported agri-food products.

However, the application of such measures caused adverse side effects over a short period of time. The high guaranteed prices naturally encouraged the increase in production (especially for wheat, butter and beef), which consecutively turned into overproduction, which led to the exponential growth of agricultural expenditure. From the moment the overproduction side effects occurred, there has been a constant struggle to reform the CAP, thus diminishing the exceeding expenditure from the European common budget through the EAGGF (Skogstad and Verdun, 2009). Additionally, the more numerous Member States became divided according to their contribution to the common budget and their benefits from the CAP, with increasing British pressure to reduce the agricultural expenditure.

Sicco Mansholt tried reforming the CAP in 1968, by land consolidation, as he considered that larger exploitations would have been more market oriented, hence enabling reduction of the protectionist measures. However, this idea did not find sufficient support at the time, and subsequently little action was taken during the 1970s and 1980s (European Institute of Romania, 2003). A first step towards reforming the CAP in order to reduce the overproduction was to introduce production quotas, especially for products that significantly exceeded the market demand, such as cereals, sugar and dairy products. These production quotas limited the producers' right to a guaranteed income based on a maximum level of production (European Institute of Romania, 2003); they were subsequently extended to all agricultural products, receiving the name of stabilizers, being designed to control Community expenditure on the common market organization. The stabilizers respected the following principle: if production exceeded a certain level (maximum guaranteed quantity), support to farmers was automatically reduced. The reduction applied to the whole production, not just the part that exceeded the maximum guaranteed quantity. For example, in the case of cereals, the maximum guaranteed quantity was set at 160 million t per year; if this ceiling was exceeded, the price was reduced by 3% during the year following that harvest (Drăgan, 2005).

However, the stabilizers have had limited success, with the accumulation of surpluses, especially for beef and milk, continuing to occur. As global negotiations for trade liberalization within the World Trade Organization progressed, Europe, together with the USA, has put pressure on the countries of the world to abandon customs barriers and allow free trade. However, in order to achieve this goal, both Europe and the USA had to make concessions on the most protected domestic economic sector – agriculture (Luca, 2009). The successful conclusion of the Uruguay Round determined the constant reduction in global agricultural prices, thus enhancing an increase in the volume of subsidies for export within the Community.

The first notable results of agricultural reformation were seen in 1992, through the reform of Ray MacSharry, European Commissioner for Agriculture at the time (European Commission, 2017g). The central element of the reform was the lowering of prices for the products generating the highest surpluses, especially cereals, beef and butter, accompanied by the payment of compensatory sums to farmers for their losses (European Institute of Romania, 2003). Moreover, compensatory payments for arable land were conditioned by setting aside a minimum of 10% of the area owned by the farmers; early retirements were also encouraged. Thus, direct payments have replaced price support. If up to that point farmers benefited from a high guaranteed price, after the reform farmers received a compensatory payment irrespective of the level of production, the reference prices being significantly reduced (Luca, 2009).

According to the European Institute of Romania (2003), this direct payments system offers, as an alternative to subsidizing agriculture, several advantages over the price subsidy:

1. The degree of transparency increases. In the system of supporting agricultural incomes by price, consumers pay high prices without knowing to what extent they subsidize agriculture. In the new system, part of the financial effort for subsidizing has passed from consumers (by reducing prices) to taxpayers through the tax system.

2. The direct payments system is beneficial for agricultural producers. In the system of supporting agriculture by price, agricultural subsidies are instead beneficial to the various intermediaries between producers and consumers, namely wholesalers, processors, storage agencies and exporters. Producers are only indirectly supported by the fact that the guarantee of the intervention price ensures their income stability, but they actually obtain the prices negotiated with the wholesalers and not the high prices on the market.

3. Incentives for overproduction are greatly diminished, on the one hand, by the lower price level that can be obtained on the market, and on the other hand, by decoupling direct payments from production volume.

4. By moving the centre of gravity of subsidy levers from markets to producers, it was expected that small farmers would benefit from subsidies proportionally with larger farms; before decoupling financial aid from production, a minority of farms – large ones, accounting for about 20% of the total – received about 80% of total aid. This situation has arisen precisely because of the system of income support through the price: those who produced more (large farms) attracted their income proportionally.

The main result of the direct payments system proposed by the MacSharry reform was that decoupling broke the vicious process of the previous decades (Fig. 2.1).

The encouraging effects of the MacSharry reform offered sufficient incentives for another reform adopted in 1999, which in essence proposed the same trend for reducing the price of agricultural products; however, this reform brought to the fore several structural components of the agricultural policy such as environmental protection, rural development and qualitative agricultural products. Falling under the Agenda 2000 measures, this new reform was based on the results and agreements obtained during the Conference held in Cork (Ireland) in 1996, where more than 5000 participants debated the future of the CAP, leading to the creation of the second pillar of the CAP (see below).

The modernization of villages and raising the standard of living for the rural population were among the initial objectives of the CAP, but the need for self-sufficiency of the Community determined an emphasis on production rather than on rural development; according to Luca (2009), the CAP confused the village with agriculture. Therefore, the new reform shifted the focus to some extent from the productive segment to rural development.

Hence, starting with the new millennium, and based on the results of the Cork Conference, this new focus became the second pillar of the CAP (European Institute of Romania, 2003) and so the CAP was no longer centred

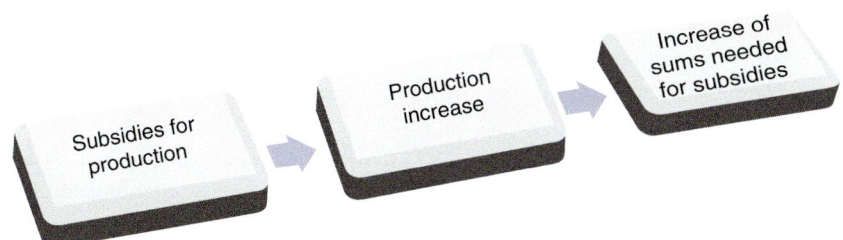

Fig. 2.1. CAP expenditure before the MacSharry reform. (Based on information from Luca, 2009.)

only on agricultural production and compensatory payments. Thus, the CAP was divided into two pillars (Luca, 2009):

- Pillar I: geared towards agriculture as an economic branch – direct payments and market interventions (namely traditional CAP).
- Pillar II: oriented towards rural development (modernization of villages and agriculture, development of alternative economic branches of agriculture, and protection of the environment and the rural landscape).

With the new second pillar, the CAP became a more complex policy but allowed more freedom for national governments to decide the steps for modernizing their rural areas, as common decisions could not be reached for such different rural areas among Member States (villages develop rather different needs in Belgium and in Spain or Greece, for example). Thus, Pillar I remained at a supranational level in terms of decisions and funds allocation, while the Member States decided more under Pillar II (European Commission, 2018a).

In 2003, another reform followed – the Fischler Reform, named after the EU's Commissioner for Agriculture, Rural Development and Fisheries between 1995 and 2004, Franz Fischler (Austria) – considered by many analysts to be the most radical in the history of the CAP (Daugbjerg and Swinbank, 2007; Swain, 2013; Swinnen, 2008). The main element of this reform was the introduction of single farm payments irrespective of the structure of production (Feichtinger and Salhofer, 2015). In addition, two new instruments have been introduced, which predict the future development of the CAP (Luca, 2009):

1. 'Cross-compliance': to receive subsidies, farmers must follow certain environmental and livestock welfare rules.
2. 'Modulation': moving funds from Pillar I – Subsidies to Pillar II – Rural Development by reducing subsidies to large farms. In other words, very large farms receive less money than they should on the basis of size, and the extra funds are moved to rural development.

The direct payments system has been significantly changed with the CAP reform of 2003. As mentioned above, the single farm payment was introduced to replace the earlier direct payment schemes. Single farm payments, in theory, allows farmers to better react to market forces, producing what is needed at a certain moment in time rather than what will give them the largest subsidy. It also gives farmers better options for managing their farms in an environmentally friendly way, as there is no need for land cultures in order to receive the payments but only to keep it in good condition (Knight, 2010). The Member States are allowed to decide the definition of a farm that could receive direct payments; however, the reformed CAP set the minimum dimension at 0.3 ha.

In contrast, Pillar II has benefited significantly from the modulation process. According to Knight (2010), before the decision to undertake modulation, the rural development segment of the CAP received only approximately 10% of the agricultural budget; this share increased substantially after funds were reallocated from the large farms of the Community towards the improvement of the rural area.

Furthermore, in 2008, the EU agriculture ministers reached a political agreement on the 'Health Check' of the CAP (European Commission, 2017b), which was meant to solve some of the issues that the 2003 reform left unsettled. One of the 2008 milestones in the CAP reform referred to national milk quotas, which were supplemented by 1% per year in order to be eliminated by 2015 (as explained previously; in the case of milk production, for example, quotas stand for the maximum amount of milk a country can produce). In terms of market purchases at the intervention price, for wheat this was limited to 3 million t (across the EU), with interventions over this quantity being made through a public auction (at a lower price). Moreover, the set-aside measure has been abolished (namely the obligation to leave some of the land uncultivated to limit the supply of products) (Luca, 2009). At the time, ministers also agreed to increase modulation, whereby direct payments to farmers are reduced and the money is transferred to the Rural Development Fund (European Commission, 2009, 2017b). This measure was meant to increase the CAP's response to the new challenges and opportunities faced by European agriculture, including climate change, the need for better water management, the protection of biodiversity and the production of green energy (European Commission, 2017b).

Although food security remains a global concern (Drăgoi *et al.*, 2018) and was the main reason for creating the CAP in the first place, the challenges that the CAP must address are related to overcoming this old paradigm, which was characteristic of post-war Europe.

In June 2013, the EU institutions agreed on a new direction for the CAP. The Cioloş reform – named after the Agriculture Commissioner at the time, Dacian Cioloş (Romania) – was modelled on a full public debate with citizens and stakeholders, with the aim of adapting it to the new challenges in terms of: (i) the competitive position of European agriculture; (ii) correctness and diversity of farming systems in Europe; (iii) climate change and the protection of natural resources; and (iv) relations between actors in the food chain (European Commission, 2014).

Basically, the post-2013 CAP is focused on ensuring a reliable source of healthy, affordable food by distributing direct payments more equitably among the Member States and among farms, thus strengthening the position of farmers in the food chain to give them the opportunity to get the best price on the market for their products and to ensure better protection against price volatility.

The European Commission has emphasized once more the need for a more efficient agricultural sector in the EU, and the new paradigm of the CAP encourages a more efficient use of natural resources to combat climate change and protect biodiversity. Overall, 30% of direct payments and 30% of rural development funds will be allocated to sustainable production methods, while specific aid will be provided for organic farming. Additionally, more funds will be allocated to related fields of work such as research, innovation and knowledge exchange, in order to encourage closer collaboration between researchers and farmers, helping them to modernize and increase their production using fewer resources for better results (European Commission, 2014).

In terms of the development of rural communities, an increased focus is based on attracting more young farmers, who would be eligible for an additional financial support during their first years of farming activities; rural tourism has also been highlighted as an opportunity for agri-businesses resilient to market shocks (Drăgoi *et al.*, 2017).

Therefore, the continuous reformation of the CAP is currently based on the two pillars that affirm the new agricultural paradigm:

- Pillar I: the common market organizations, which include common measures to regulate the functioning of the integrated markets of the agricultural products.
- Pillar II: rural development, which includes structural measures aimed at promoting the development of rural areas.

Table 2.1 highlights the most important moments in the development of the CAP since 1957.

As the following sections will highlight, despite its massive political and financial weight, agriculture is a small part of the EU economy, accounting for only 1.7% of gross domestic product (GDP) and employing 4.6% of the workforce, according to Organisation for Economic Co-operation and Development (OECD) figures (EurActiv, 2017). However, the EU and the USA are in competition for the title of largest exporter of agricultural products, and the value of EU exports reached €130.7 billion in 2016; hence, significant importance is still attributed to the CAP.

Consequently, discussions regarding the evolution of the CAP after 2020 are already taking place, and the central element agreed on by all stakeholders is that the policy needs urgent simplification, without renouncing its wider objectives. The current European Commissioner for Agriculture and Rural Development, Phil Hogan, sustains that 'where appropriate, greater subsidiarity for Member States would allow for sufficient flexibility to manage the policy in the most appropriate manner, while the EU continues to set the objectives and targets to be achieved'.

The evolution of the CAP's measures through continuous reformation is also visible when analysing the expenditure of this sector from the total budget of the EU. Figure 2.2 shows how, over time, the reforms have managed to decrease agricultural expenditure and decouple financial incentives for farmers from agricultural production.

According to a European Commission report (European Commission, 2013), in 1992, market management represented over 90% of total CAP expenditure, consisting of export refunds and intervention purchases. By the end of 2013, it had dropped to just 5% as market intervention has become a safety-net tool for times of crisis and direct payments are the major source of support, 94% of which are decoupled from production. Furthermore, since 2014, the share of expenditure between the two pillars of the CAP has been open to the possibility of transferring up to 15% of the national envelopes of the Member States between pillars, enabling them to adapt spending and allocations to their national priorities.

Tables 2.2 and 2.3 provide information based on European Commission reports regarding the CAP allocations for each Member State for the current multi-annual financial framework 2014–2020. As expected, the main beneficiaries of this policy are some of the old Member States, such as France, Italy and Germany, along with the Mediterranean countries. As highlighted previously, agricultural expenditure represented a large consumer of the EU's budget, especially before the reformation process began. Therefore, the Member States

Table 2.1. Key moments in the evolution of the CAP. (Based on information from European Commission, 2012, 2017g.)

Year	Main changes
1957	The Treaty of Rome was signed creating the European Economic Community among six founding European countries as a basis for the future common market of the EU. The CAP was foreseen as a common policy with the objectives of providing affordable food for EU citizens and a fair standard of living for farmers.
1962	The CAP was created with the main goal of ensuring a sufficient food supply for the population of the Member States.
1970s	This period was characterized by overproduction and excessive expenditure in order to sustain surpluses. The Community faced 'food mountains' and 'milk lakes', namely excessive agricultural products that surpassed the Community's demands and needs.
1984	The Community introduced milk quotas, and by the end of the 1980s, production quotas were extended to all products, being named stabilizers.
1992	The MacSharry reform (named after the European Commissioner in charge at the time with the portfolio of agriculture) was the first major reform of the CAP and marked a shift from market support to producer support, as farmers were encouraged to be more environmentally friendly.
1999	Under the Agenda 2000 document, the new focus of the CAP was rural development along with attempts to align the food supply to market demand. New emphasis was put on developing rural areas of the EU from an economic and social perspective, with rural development becoming the second pillar of the CAP.
2003	A new reform that cut links between production and subsidies was adopted; to receive an income aid, farmers had to comply with food safety and environmental and animal welfare standards. Apart from emphasizing again the necessity of further 'decoupling' (financial support from production) and 'cross-compliance' (complying with environmental and animal welfare rules), this reform brought the concept of 'modulation' (shifting funds from subsidies to rural development).
Mid-2000s	The CAP was opened up to the less developed countries of the world, thus becoming the largest food importer from these countries.
2007	The enlargement of the EU increased the number of farmers (and the number of small, family farms) and addressed new issues that the reformed CAP had to tackle.
2008	The 'Health Check' of the CAP proposed abolishing the production quotas for milk and abolishment of the set-aside previous rules, along with reducing financial aid for large farms.
Post-2013	The new CAP paradigm included environmental protection and the fight against climate change, while ensuring agricultural employment, especially in the case of the young population, rural development and ecological products for the consumers.

were divided in terms of agricultural beneficiaries or contributors, and the main beneficiaries who supported the CAP during its first decades can still be found among the beneficiaries today. The excessive spending for the CAP represented one of the main reasons for negotiating and obtaining the British rebate in

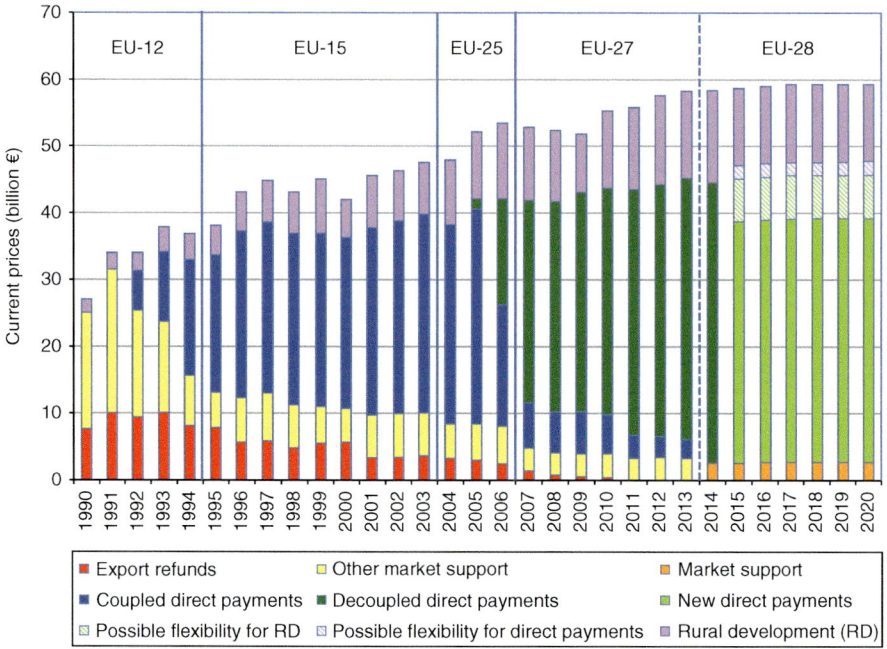

Fig. 2.2. CAP expenditure by year (in current prices). (From European Commission, 2013.)

1984, and decades later, it became one of the motivations for the supporters of the Brexit vote[3].

As the EU faces new challenges in terms of internal economic and social integration, the exit of the British population, economic constraints regarding the Euro area, developments of the global agricultural market, climate change threats and so on, one of the oldest common policies of the EU – the CAP – needs future improvement in order to face and answer these challenges. Thus, the new paradigm of the CAP provides new lines of action for modernization of the agricultural sector.

One of the most recent communications of the European Commission (European Commission, 2017f) outlines suggestions on the future of food and farming by listing a set of new measures to be taken in order to simplify the CAP. In terms of climate change, the main issue that the new CAP must face is to provide protection for farmers against extreme weather conditions and climate change, while also ensuring that the European farming sector itself does not aggravate the problem of climate change; subsequently, a more sustainable and ecological agricultural sector must be developed over the following decades, as the EU prepares to welcome its fouth generation of farmers. Each Member State must develop a strategic plan, according to the national specificity, in order to support the EU to meet the climate standards agreed at the international level. Rather than on compliance, the attention on the Member States will be paid more on monitoring progress and ensuring funding is focused on concrete results (European Commission, 2017f).

Table 2.2. Direct payments (current prices in €1000): ceilings by Member State. (From European Commission, 2017c.)

Member State	Financial year						Total 2015–2020
	2015	2016	2017	2018	2019	2020	
Austria	693,716	693,065	692,421	691,754	691,746	691,738	4,154,440
Belgium	544,047	536,076	528,124	520,170	512,718	505,266	3,146,401
Bulgaria	642,103	721,251	792,449	793,226	794,759	796,292	4,540,080
Croatia	113,908	130,550	149,200	186,500	223,800	261,100	1,065,058
Cyprus	51,344	50,784	50,225	49,666	49,155	48,643	299,817
Czech Republic	875,305	874,484	873,671	872,830	872,819	872,809	5,241,918
Denmark	926,075	916,580	907,108	897,625	889,004	880,384	5,416,776
Estonia	110,018	121,870	133,701	145,504	157,435	169,366	837,894
Finland	523,247	523,333	523,422	523,493	524,062	524,631	3,142,188
France	7,586,341	7,553,677	7,521,123	7,488,380	7,462,790	7,437,200	45,049,511
Germany	5,178,178	5,144,264	5,110,446	5,076,522	5,047,458	5,018,395	30,575,263
Greece	2,047,187	2,039,122	2,015,116	1,991,083	1,969,129	1,947,177	12,008,814
Hungary	1,272,786	1,271,593	1,270,410	1,269,187	1,269,172	1,269,158	7,622,306
Ireland	1,216,547	1,215,003	1,213,470	1,211,899	1,211,482	1,211,066	7,279,467
Italy	3,953,394	3,902,039	3,850,805	3,799,540	3,751,937	3,704,337	22,962,052
Latvia	168,886	195,649	222,363	249,020	275,887	302,754	1,414,559
Lithuania	393,226	417,890	442,510	467,070	492,049	517,028	2,729,773
Luxembourg	33,662	33,603	33,545	33,486	33,459	33,431	201,186
Malta	5,240	5,127	5,015	4,904	4,797	4,689	29,772
Netherlands	793,319	780,815	768,340	755,862	744,116	732,370	4,574,822
Poland	2,970,020	2,987,267	3,004,501	3,021,602	3,041,560	3,061,518	18,086,468
Portugal	557,667	565,816	573,954	582,057	590,706	599,355	3,469,555
Romania	1,428,531	1,629,889	1,813,795	1,842,446	1,872,821	1,903,195	10,490,677
Slovakia	377,419	380,680	383,938	387,177	390,781	394,385	2,314,380
Slovenia	138,980	137,987	136,997	136,003	135,141	134,278	819,386
Spain	4,833,647	4,842,658	4,851,682	4,866,665	4,880,049	4,893,433	29,168,134
Sweden	696,487	696,890	697,295	697,678	698,723	699,768	4,186,841
UK	3,548,576	3,555,915	3,563,262	3,570,477	3,581,080	3,591,683	21,410,993
Total EU-28	41,679,856	41,923,877	42,128,888	42,131,826	42,168,635	42,205,449	252,238,531

Table 2.3. Breakdown of Union support for rural development 2014–2020 (current prices in €1000). (From European Commission, 2017c.)

	2014	2015	2016	2017	2018	2019	2020	Total 2014–2020
Austria	557,806	559,329	560,883	562,467	564,084	565,713	567,266	3,937,551
Belgium	78,342	78,499	78,660	78,824	78,991	79,158	79,314	551,790
Bulgaria	335,499	335,057	334,607	334,147	333,680	333,187	332,604	2,338,783
Croatia	332,167	332,167	332,167	332,167	332,167	332,167	332,167	2,325,172
Cyprus	18,895	18,893	18,891	18,888	18,886	18,883	18,875	132,214
Czech Republic	314,349	312,969	311,560	310,124	308,659	307,149	305,522	2,170,333
Denmark	90,287	90,168	90,047	89,924	89,798	89,665	89,508	629,400
Estonia	103,626	103,651	103,676	103,702	103,728	103,751	103,751	725,886
Finland	335,440	336,933	338,456	340,009	341,593	343,198	344,776	2,380,408
France	1,404,875	1,408,287	1,411,769	1,415,324	1,418,941	1,422,813	1,427,718	9,909,731
Germany	1,178,778	1,177,251	1,175,693	1,174,103	1,172,483	1,170,778	1,168,760	8,217,851
Greece	601,051	600,533	600,004	599,465	598,915	598,337	597,652	4,195,960
Hungary	495,668	495,016	494,351	493,672	492,981	492,253	491,391	3,455,336
Ireland	313,148	313,059	312,967	312,874	312,779	312,669	312,485	2,189,985
Italy	1,480,213	1,483,373	1,486,595	1,489,882	1,493,236	1,496,609	1,499,799	10,429,710
Latvia	138,327	138,361	138,396	138,431	138,467	138,498	138,499	968,981
Lithuania	230,392	230,412	230,431	230,451	230,472	230,483	230,443	1,613,088
Luxembourg	14,226	14,272	14,318	14,366	14,415	14,464	14,511	100,574
Malta	13,880	13,965	14,051	14,139	14,230	14,321	14,412	99,000
Netherlands	87,118	87,003	86,886	86,767	86,645	86,517	86,366	607,305
Poland	1,569,517	1,567,453	1,565,347	1,563,197	1,561,008	1,558,702	1,555,975	10,941,201
Portugal	577,031	577,895	578,775	579,674	580,591	581,504	582,317	4,057,788
Romania	1,149,848	1,148,336	1,146,793	1,145,218	1,143,614	1,141,925	1,139,927	8,015,663
Slovakia	271,154	270,797	270,434	270,062	269,684	269,286	268,814	1,890,234
Slovenia	118,678	119,006	119,342	119,684	120,033	120,384	120,720	837,849
Spain	1,187,488	1,186,425	1,185,344	1,184,244	1,183,112	1,182,137	1,182,076	8,290,828
Sweden	248,858	249,014	249,173	249,336	249,502	249,660	249,768	1,745,315
UK	371,473	370,520	369,548	368,557	367,544	366,577	365,935	2,580,157
Total EU-28	13,618,149	13,618,658	13,619,178	13,619,708	13,620,249	13,620,801	13,621,363	95,338,109

With the purpose of further reducing agricultural costs, especially as the UK's contribution to the EU budget will become unavailable after 2020, even though the support for farmers will continue through the system of direct payments, the new CAP must ensure that these payments are directed towards farmers who need it the most, hence guaranteeing a more efficient policy with more visible results. Therefore, the future CAP will encourage the use of modern technologies, as opposed to the extensive agriculture of the mid-1950s and 1960s. Additionally, as farming could again represent a sustainable form of employment, as it was in previous decades, the new CAP aims at encouraging young people to remain in rural areas and work in the agricultural sector.

Another important aspect that arose at the beginning of the millennium is ecological agriculture, with the EU assuming an important role in providing safe food for its citizens, while also ensuring proper living, feeding and breeding conditions for livestock. Furthermore, the CAP aims to offer useful tools on risk management to help farmers cope with the uncertainty of climate, market volatility and other risks (European Commission, 2017f).

Under the new paradigm, the EU proposes a new concept regarding 'smarter villages', where traditional and new networks and services are enhanced by means of digital and telecommunication technologies and innovations and better use of knowledge (European Commission, 2017a).

Acknowledging that the CAP continues to be the most important EU policy intervening in the EU rural economy in terms of funding and the range of instruments employed, the measures suggested under the rural development pillar target rural business development, including the modernization of farms, investments in small-scale local infrastructure and connectivity projects, village renewal, knowledge development, knowledge sharing, and bottom-up initiatives (European Commission, 2017a).

However, until drones destroying weeds, satellite-supervised crops and robots involved in the wine-making process (European Commission, 2017e) become available in all European rural regions, smaller steps towards rural development can be achieved with the main goal of providing social inclusion, diminishing inequities and reducing the unemployment, particularly youth unemployment, rates. One such example that best exemplifies the need for better integrated policies across the EU is the agritourism sector, which provides one viable option for present and future progress of non-urban areas (Drăgoi *et al.*, 2017). According to the authors, the main benefits that business ventures in agritourism may induce include but are not limited to:

- increased revenues and employment opportunities in rural regions;
- protection of the natural environment by appealing to traditional and non-invasive practices;
- promotion of local culture and heritage;
- consolidation of the entrepreneurial skills and personal objectives of the involved population, as an alternative solution to the decreasing agricultural endeavours in certain regions.

There is no doubt that the new agricultural paradigm embarked on by the EU emphasizes new problems that the CAP (along with other European policies, such as the Cohesion Policy) must tackle and also indicates the need for modern tools to meet these goals. The agriculture of the 1960s has certainly changed significantly from farming nowadays analysed as a whole; nevertheless, when addressing the agricultural sector in a more detailed manner, there are a noteworthy number of discrepancies observed at national and regional levels among the Member States.

Given that the post-2014 CAP allows each Member State to choose the direct payment scheme that suits best their national farmers' needs, while allowing a new recoupling with production, the new policy paradigm permits a certain degree of decentralization but could also accentuate national gaps. In this context, the following section provides a more detailed description of the EU's agricultural sector.

2.2   The Agricultural Sector in the EU Member States

The EU has developed substantially, from six Member States to 28[4] in the past decades. Since creation of the CAP in 1962, the Community has been divided, as described above, into Member States supporting the CAP objectives and means of implementation and Member States against the CAP regulations. If we look at the EU map comprising the current Member States (Fig. 2.3), which at various moments in time were candidate countries, it is easily understood why the CAP determined such segregation.

As Fig. 2.3 shows, the location and the size of each Member State vary considerably, determining the variable share of their territories used for agricultural purposes. While some countries benefit from a temperate or Mediterranean climate for instance, others have colder temperatures, longer winters and less sunshine throughout the year. At the same time, the area of some countries and their populations are small, such as Malta, Cyprus, Luxembourg and the Baltic countries compared with Germany, France, the UK and Spain.

The national land used for agriculture varies among the Member States. Eurostat data (Eurostat, 2015b) shows that there are discrepancies in terms of the percentage of national land employed in agricultural activities (Fig. 2.4). Agricultural land use is the most common primary land-use category in the EU-27 (Croatia joined in 2013 to complete the EU-28) as presented in Fig. 2.4, accounting for 43.5% of the total area in 2012. According to Eurostat (2015b), land in agricultural use encompasses various land cover types: the most common are arable land, permanent crops and grassland, and in 13 out of the 27 EU Member States taken into analysis at that time, more than half of the land area was used for agricultural purposes. The highest share of agricultural land was recorded in Ireland (71.5%), while Denmark, the UK, Hungary and Romania each reported shares of more than 60.0%. On the opposite pole, Finland and Sweden's agriculture played a minor role in terms of land use, accounting for less than 10.0% of the total land area in these two Member States (Eurostat, 2015b). The main explanation for the fact that these northern countries employ little land in agriculture is that most of their territory is used for forestry purposes, offering additional land-use activities such as hunting,

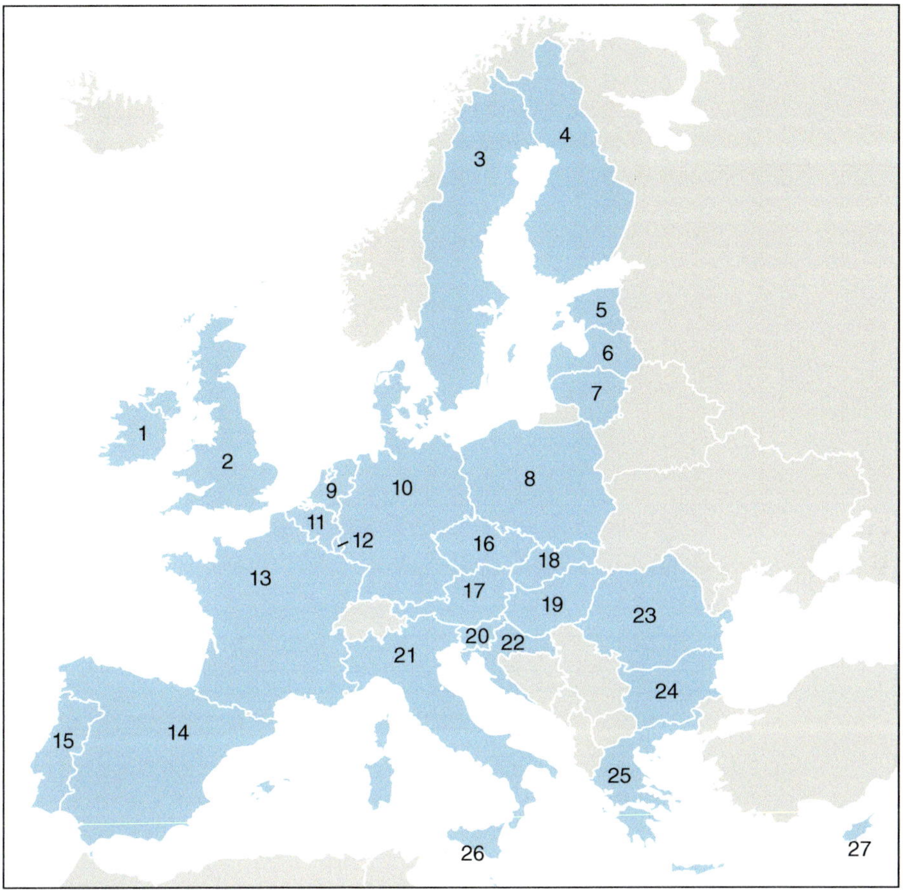

1. Ireland	12. Luxembourg	25. Greece
2. UK	13. France	26. Malta
3. Sweden	14. Spain	27. Cyprus
4. Finland	15. Portugal	
5. Estonia	16. Czech Republic	
6. Latvia	17. Austria	
7. Lithuania	18. Slovakia	
8. Poland	19. Hungary	
9. Netherlands	20. Slovenia	
10. Germany	21. Italy	
11. Belgium	22. Croatia	
	23. Romania	
	24. Bulgaria	

Fig. 2.3. Map of the EU (28 Member States). (Public domain via Wikimedia commons.)

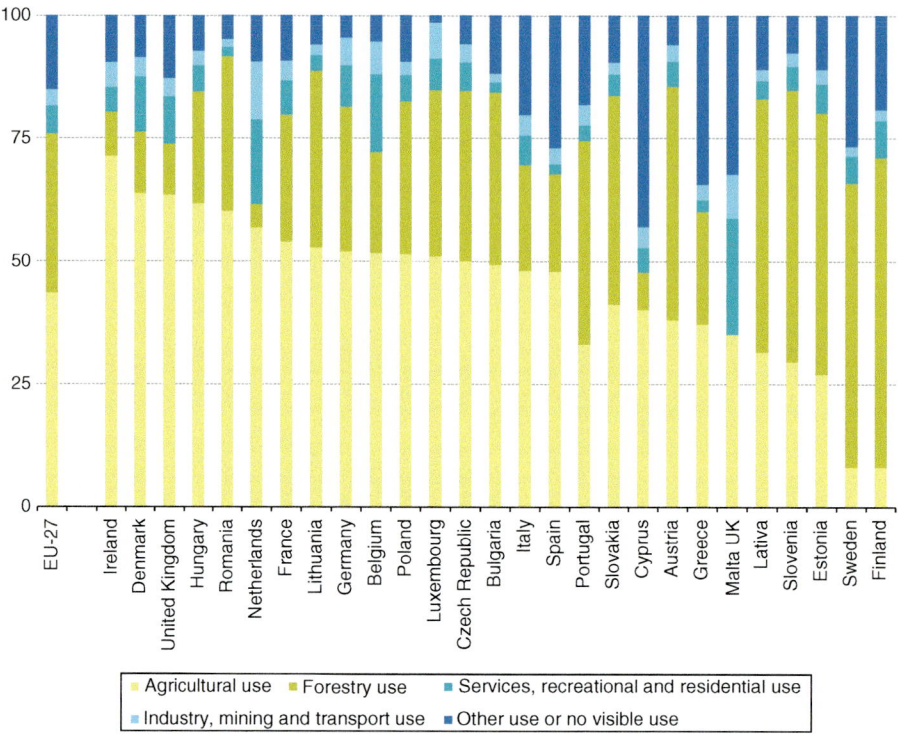

Fig. 2.4. Primary land use by land-use type in 2012 (% of total area) of the EU-27. Forestry use not available for Malta. (From Eurostat, 2015b.)

bird and animal watching, and hiking. Their position on the European map, as seen in Fig. 2.3, also determines a less extensive agricultural sector, as their location leads to a lack of sun and low temperatures, which are inadequate for a varied plant crop production.

Arable land (which includes land for cereals and other arable land) accounted for 59.8% of the utilized agricultural area in the EU-28 in 2013 (Fig. 2.5). Permanent grassland and meadow (which is composed of pasture, meadow and rough grazing) accounted for just over one-third (34.2%). Vineyards, olive trees and orchards, which represent the permanent crops, accounted for a 5.9% share, with the remaining 0.2% attributed mainly to kitchen gardens (Eurostat, 2015a).

The relative importance of arable land varies considerably among EU Member States, and its share of the agricultural area used ranged from 35.6% in Slovenia to 81.6% in Hungary and higher in the Nordic Member States of Sweden (85.1%), Denmark (91.5%) and Finland (98.5%), while Portugal (30.2%) and Ireland (21.0%) had lower values (Eurostat, 2015a).

Figure 2.5 also shows that the share of meadow and grassland surpassed 50% of the agricultural area in countries such as Luxembourg, Slovenia and the UK, reaching almost 80% in Ireland. In contrast, Malta, Cyprus and Finland have less than 2% of agricultural area for permanent meadow and grassland. This

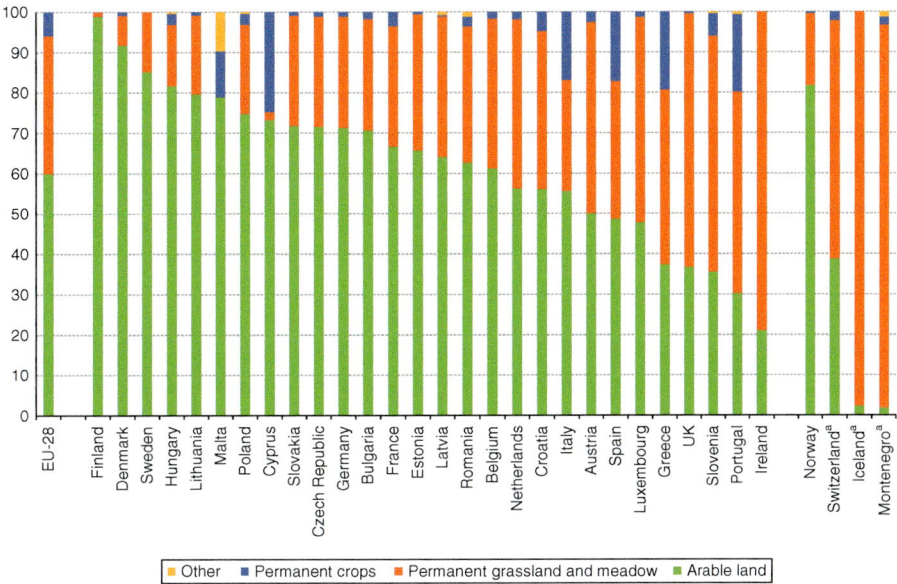

Fig. 2.5. Distribution of agricultural area use, 2013 (%). [a]2010 data. (From Eurostat, 2015a.)

analysis is highly relevant because meadows are associated with livestock rearing, especially with the production of dairy and sheep farming.

It is no surprise that permanent crops associated with orchards, such as olives and citrus fruits, have a higher share of the agricultural area in countries such as Cyprus, Malta, Greece, Spain, Portugal and Italy, as they all are southern countries renowned for their exports of such agricultural products.

Taking the analysis further, Fig. 2.6 shows that France and Spain accounted for the largest share of EU-28 agricultural land, allowing a preliminary deduction that these countries are among the most fervent supporters of the CAP. The agricultural land owned by these countries and the high representativeness they have within European institutions determines their powerful influence on the CAP decision process.

However, Fig. 2.6 allows other interesting comparisons among Member States in terms of agricultural use of land and labour force. The number of agricultural holdings records the highest value in Romania (3.6 million holdings), showing that this country has one-third of the total number of agricultural holdings of the EU-28; the UK, France and Germany lag behind at a considerable distance, with less than 5% of the total number of holdings of the EU-28. These disproportionate shares in the total number of agricultural holdings indicate the differences in the dimensions of such holdings and their relative economic efficiency in the different countries. On the one hand, countries like Romania and Poland with a medium agricultural area but with a significant number of agricultural holdings are characterized by the small sizes of these holdings. The more numerous the agricultural holdings are in one country, the smaller the number of hectares per holding they employ. Subsequently, smaller agricultural

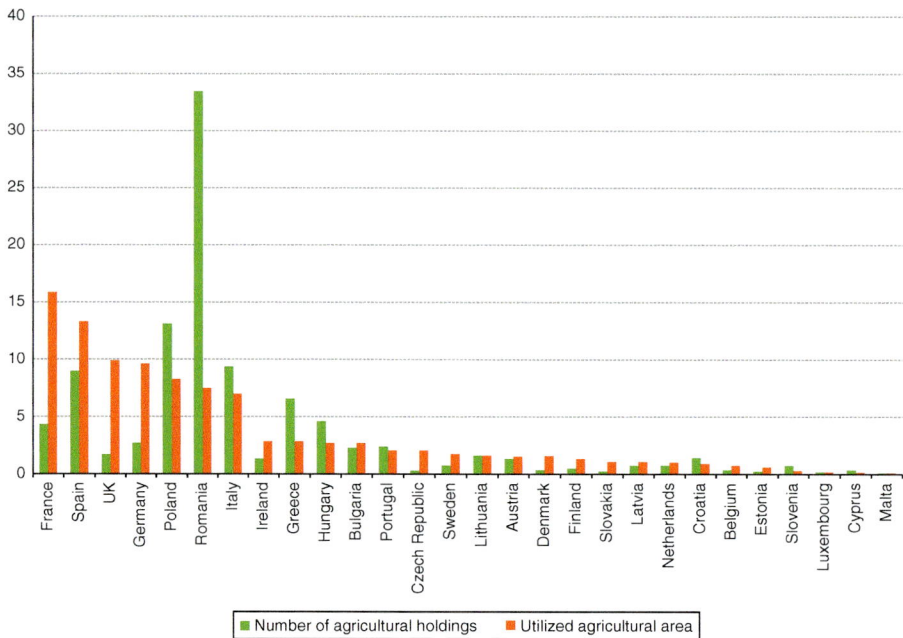

Fig. 2.6. Agricultural indicators: share of EU-28 total, 2013 (%). (From Eurostat, 2015a.)

holdings also have reduced agricultural output. On the other hand, countries such as the UK, Germany and the Czech Republic account for larger holdings with better use of the available agricultural land.

According to Eurostat (Fig. 2.7), the largest agricultural holdings are found in the Czech Republic, followed by the UK and Slovakia. At the other end of the scale, Romania ranks among the last countries along with Cyprus and Malta – which are comparably smaller than Romania, in terms of total surface and agricultural land – reporting average agricultural holding sizes well below 10.0 ha per holding. Consequently, the output of the small holdings is noticeably less than the output of large agricultural holdings of more than 90 ha found in the UK or the Czech Republic.

In fact, a strong characteristic of European very small farms is that they are subsistence households and their owners need to put in much effort to make a living (Eurostat, 2016b). Therefore, in 2013 the EU-28 recorded almost 75% very small farms (in economic terms), which consume more than half of their own production, while 42.6% of small farms were classified as subsistent. Unsurprisingly, the highest proportion of more than 90% of subsistence households among all very small farms was found in Latvia, Romania and Slovenia.

Assessing previous literature, Davidova *et al.* (2012) provide a classification of subsistence and semi-subsistence farms in terms of physical measures, economic size and market participation. Physical measures define small farms as those operating on an agricultural area of 5 ha or less; economic size is mainly applied for statistics purposes within the EU, expressed in terms of European

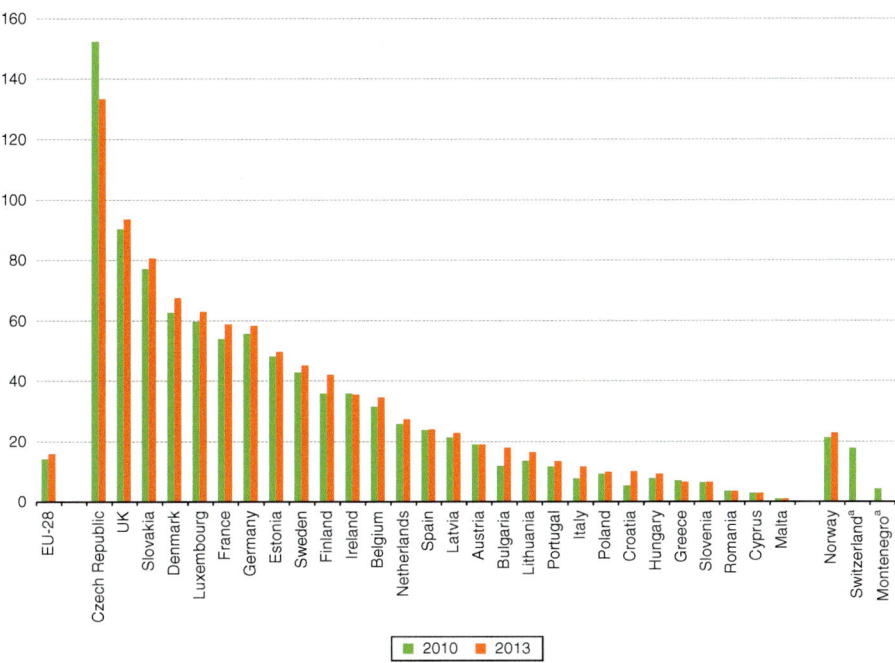

Fig. 2.7. Average agricultural area used per holding in 2010 and 2013 (hectares). Data for Iceland are not shown for reasons of scale: the 2010 value was 616 ha. [a]2013 data not available. (From Eurostat, 2015a.)

Size Units (ESU)[5]. According to the authors, farms between 1 and 8 ESU are classified as semi-subsistence. Lastly, in terms of market participation, based on Wharton's (1969) standpoint, farmers who sell less than 50% of their output are included in the semi-subsistence category, while those above this threshold can be labelled as semi-commercial or even commercial.

The joint results of previous studies (Davidova *et al.*, 2012; Galluzzo, 2017; Giurca, 2008) emphasize that, mainly in the new Member States of the EU, subsistence and semi-subsistence farmers struggle to provide sufficient agricultural output just to make a living for family members, while market participation accounts for an insufficient share to afford further investments in agricultural or agritourism activities.

Although the standard output of farms in the EU increased by almost 56% between 2005 and 2013, as the Eurostat (2016b) report shows, there were significant differences among agricultural holdings of the EU-28 Member States. The Netherlands recorded the largest farms, in economic terms, generating an average of €303,800 of standard output, derived in particular from growing high-value products such as flowers and vegetables. Except for farms in the Netherlands, Denmark, Belgium, the Czech Republic, Germany, Luxembourg, France and the UK, the rest of the Member States recorded an economic size of farms of less than €80,000 per farm. Romania is situated at the opposite end of the scale, where farms averaged €3300 of standard output. Comparing the results for the Netherlands with those for Romania, the average economic size of farms in the former was approximately 92 times larger than those in the latter. At Nomenclature of Territorial Units for Statistics (NUTS) 2 level regions, there were ten regions in the EU-28 where farms on average generated €5000 or less of standard output in 2013. All eight of the Romanian regions figured in this list, along with the Greek island region of Ionia Nisia and the Polish region of Podkarpackie. The region with the lowest level of standard output per farm (€2600) was Sud-Vest Oltenia in Romania for the year 2013 (Eurostat, 2016b).

However, the contrast goes even deeper. Romania has the largest share of under-35-year-old managers for large agricultural holdings (even though the number of such holdings is not as significant as in the case of other EU Member States).

In 2013, more than 35% of EU-28 large-farm managers were aged 45–54 years. However, as shown in Fig. 2.8, the most notable exception was in the case of Romania, where a high proportion of managers of very large farms were aged 35 years or less (more than 57%), which was more than six times higher than the EU average (approximately 9%). For other European countries, the groups aged 35–44 years and 45–54 years accounted for the largest share among large-farm managers, while in the Czech Republic, Hungary and Slovakia the commonest age of managers was 55–64 years.

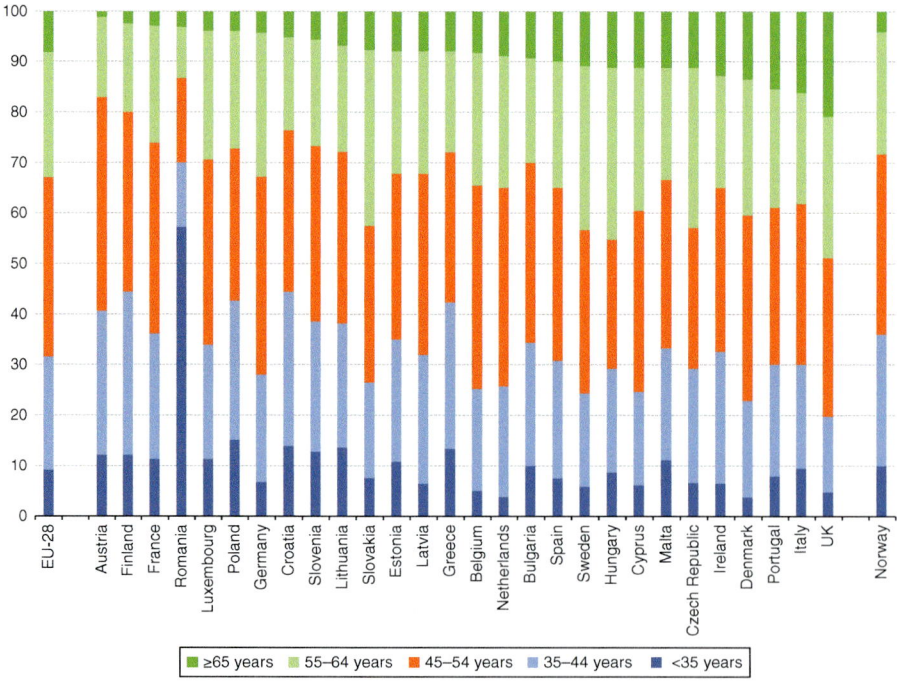

Fig. 2.8. Age of farm managers on very large farms (defined in economic terms as those with €100,000 or more of standard output), 2013 (% of total). Ranked on the share of those aged 65 years or more. (From Eurostat, 2016b.)

In contrast, in terms of small and very small agricultural holdings, Romania, where most of the farms fall into these two categories (along with Portugal, Cyprus and Bulgaria), has a high proportion (more than 40%) of managers aged 65 years and above, while the EU average is approximately 35% for this age category (Fig. 2.9).

In terms of the labour force involved in agriculture in general, the total farm labour force in the EU-28 was the equivalent of 9.5 million annual work units (AWUs)[6] in 2013, of which 8.7 million (92%) were regular workers (Fig. 2.10) (Eurostat, 2015a). Thus, corroborating the information emphasized in previous figures, the countries with larger agricultural holdings account for a higher number of regular workers in agriculture, with examples being the Czech Republic, the UK, Luxembourg, Denmark, France and Germany.

Again, Romania ranks last in terms of full-time regular workers in agriculture, with less than 8% being enrolled as full-time workers. The remainder of agricultural workers are registered as regular but not employed full time in agriculture; the small dimensions of the agricultural holdings in this country demonstrate their low capacity to employ full-time workers in agriculture,

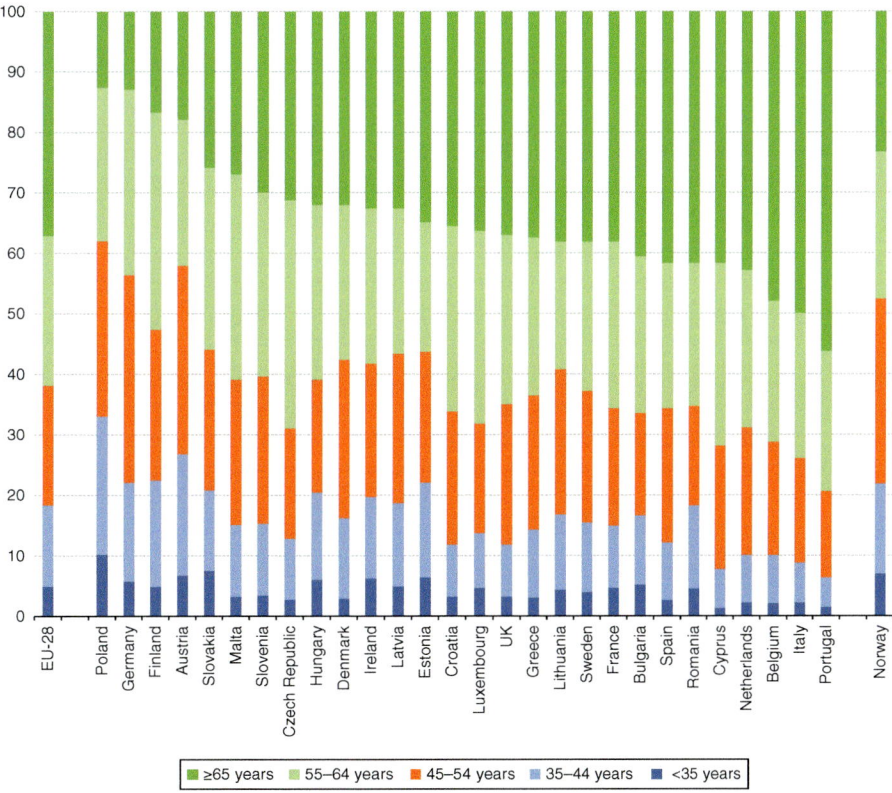

Fig. 2.9. Age of farm managers on very small and small farms (defined in economic terms as those with less than €8000 of standard output), 2013 (% of total). Ranked on the share of those aged 65 years or more. (From Eurostat, 2016b.)

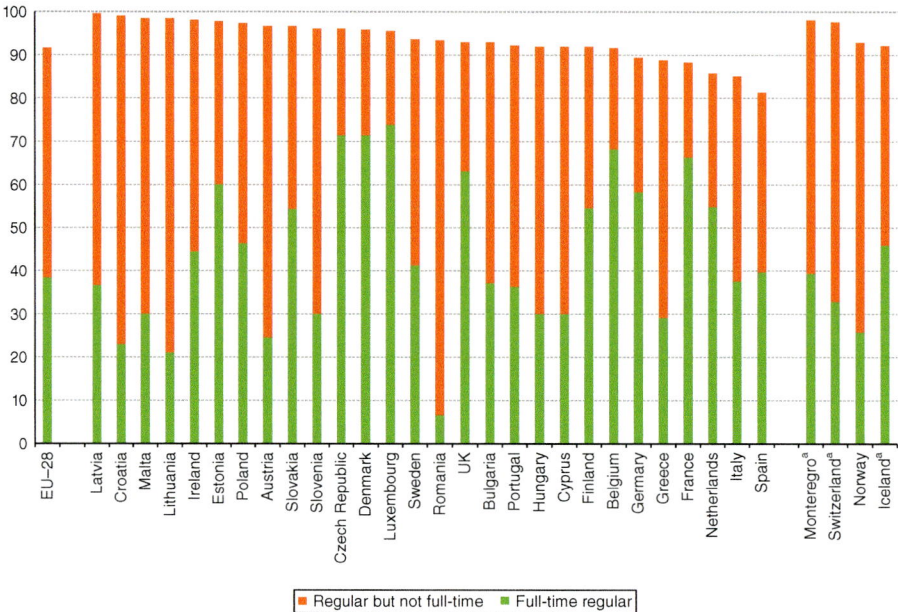

Fig. 2.10. Share of regular workers in the farm labour force and distribution by work intensity, 2013 (%). [a]2010 data. (From Eurostat, 2015a.)

in contrast to the Czech Republic, Denmark and Luxembourg, where more than 70% of agricultural workers are employed full time. Moreover, the low employment capacity and the reduced level of agricultural output also become an incentive for seeking other means of income in order to ensure a daily living, which may explain the reduced number of full-time employees in agricultural holdings in Romania. This assumption is also demonstrated in Table 2.4, where a decrease in agricultural labour input can be seen throughout almost all Member States but with an accentuated drop in the case of Bulgaria, Romania and the Baltic countries, which also account for small agricultural holdings.

As seen previously, the majority of the EU's agricultural holdings are relatively small, and in the case of several countries, such as Romania, Bulgaria and the southern states, these are usually family-run holdings (Eurostat, 2016a). Moreover, the agricultural sector in these countries is characterized by a high dependency on climate factors, with extensive production during the years with predictable weather conditions and lower production in periods of heavy rain falls and flooding, extremely low or high temperatures during mid-seasons, etc. This is why they are characterized by seasonal labour peaks, with high numbers of workers hired for relatively short periods of time, thus explaining the fact that some farmers are occupied on a part-time basis (and may seek to have alternative, sometimes significant, sources of income) (Eurostat, 2016a).

For the year 2016, EU-28 agricultural labour input was estimated at almost 9.5 million AWUs (representing the equivalent of 9.5 million people working full-time) (Table 2.4). Among all 28 Member States, the highest levels of agricultural labour input were recorded for Poland (approx. 1.6 million AWUs),

Table 2.4. Agricultural labour input in 2007, 2010 and 2013–2015 (×1000 AWUs). (Based on information from Eurostat, 2017a.)

Country/region	2007	2010	2013	2014	2015	2016	Change 2007/2016
EU-28	11,877.61	10,345.29	9,913.48	9,737.40	9,504.27	9,490.12	−20.1%
EU-27	11,668.61	10,143.29	9,722.48	9,549.40	9,322.27	9,316.12	−20.2%
Austria	136.01	127.94	124.23	121.83	119.61	117.96	−13.3%
Belgium	66.00	61.90	57.90	57.29	56.77	57.49	−12.9%
Bulgaria	494.40	406.50	321.20	297.50	276.40	256.80	−48.1%
Croatia	209.00	202.00	191.00	188.00	182.00	174.00	−16.7%
Cyprus	25.90	25.40	25.60	25.03	17.74	20.85	−19.5%
Czech Republic	127.00	108.80	105.10	104.90	104.80	104.50	−17.7%
Denmark	58.80	54.20	52.71	54.09	55.06	53.97	−8.2%
Estonia	32.92	25.36	22.27	21.97	20.28	20.30	−38.3%
Finland	90.90	82.10	75.90	81.20	79.40	71.00	−21.9%
France	867.10	809.10	781.00	774.50	761.60	752.90	−13.2%
Germany	554.20	522.00	503.00	504.00	496.00	480.00	−13.4%
Greece	574.80	441.45	467.00	454.50	442.40	430.75	−25.1%
Hungary	459.29	444.16	444.42	462.93	441.90	434.28	−5.4%
Iceland	4.57	4.21	3.98	3.50	3.80	3.78	−17.4%
Ireland	150.20	165.60	163.64	163.64	163.64	163.64	8.9%
Italy	1,212.00	1,164.00	1,077.50	1,095.30	1,111.80	1,125.30	−7.2%
Latvia	107.37	85.88	82.87	76.42	77.87	76.25	−29.0%
Lithuania	158.00	143.40	144.80	149.90	150.80	148.80	−5.8%
Luxembourg	3.78	3.73	3.56	3.53	3.53	3.45	−8.7%
Malta	4.20	4.90	5.00	5.00	5.00	5.00	19.0%
Netherlands	157.60	150.40	148.20	146.20	146.50	146.40	−7.1%
Norway	61.40	51.40	48.10	47.00	45.90	45.00	−26.7%
Poland	2,299.30	1,914.80	1,937.10	1,937.10	1,937.10	1,675.80	−27.1%
Portugal	351.30	309.41	281.33	265.15	258.18	250.72	−28.6%
Romania	2,205.00	1,639.00	1,564.00	1,433.00	1,293.00	1,592.00	−27.8%
Slovakia	91.30	56.10	54.20	53.90	48.90	48.70	−46.7%
Slovenia	83.95	77.01	82.75	81.80	81.37	79.97	−4.7%
Spain	998.23	963.77	841.68	824.28	818.74	849.23	−14.9%
Sweden	68.60	65.30	62.08	60.80	59.55	58.33	−15.0%
Switzerland	86.38	80.74	77.70	77.36	76.49	75.64	−12.4%
UK	290.46	291.09	293.45	293.64	294.32	291.73	0.4%

Romania (almost 1.6 million AWUs) and Italy (approx. 1.1 million AWUs), even though the areas and populations of these countries are not quite comparable.

Over the past decade, there has been a significant reduction in agricultural labour input in the EU-28 of approximately one-fifth; the steepest declines were found in Bulgaria (48% reduction in 2016, compared to 2007), Slovakia (−46.7%) and Estonia (−38.3%). Only three Member States recorded an increase during the same period: Malta (19%), Ireland (8.9%) and the UK (0.4%).

Agricultural income, as a key measure for determining the viability of the agricultural sector and expressed as an index (2010 being the basis of this index), is a measure of relative labour productivity (Eurostat, 2016a). This indicator must be linked to the total labour force input in agriculture: the increase in agricultural income is based on the diminished level of AWU input in agriculture, resulting in the income being shared among a smaller workforce. Table 2.5 highlights that a group of 11 Member States showed a decrease in agricultural income per AWU in 2016, compared with 2010, the steepest decreases being recorded in Denmark, Estonia and Malta. In Malta, this decrease could be attributed to the increase in total AWUs in agriculture, which rose by 19% in 2016, compared with 2007, as shown in Table 2.4. For the other countries, a possible explanation resides in the amplified value of commodity prices, thus determining the decrease in income in general, not just in the agricultural sector.

The index of agricultural income per AWU rose in the remaining EU Member States between 2010 and 2016, from barely significant percentages in the Netherlands (101.0) and Luxembourg (103.6) to more extensive differences in the case of Bulgaria (188.2) and the Czech Republic (155.1).

On the other hand, the agricultural output value in 2016 was more than €405 billion, with the highest share for vegetables and horticultural plants (13.2%), milk (12.2%) and cereals (10.7%), as shown in Fig. 2.11. Analysing these results in more depth, a higher share of crops in the total agricultural output (almost 52%) compared with livestock (39.2%) can be observed as an overall tendency during the past decades.

This upward trend of crop production is also explained by observed pattern shifts in overall diet preferences. The vegetarian tendencies in Europe (Drăgoi, 2016) and alarm signals from the World Health Organization regarding increasing obesity rates (World Health Organization, 2017) have determined an amplification of crop production compared with animal production. Additionally, ecological agriculture has come to play a growing role in the European population's diet preferences; thus, organic plant products are likely to experience a greater share of total plant crop production in the following decades.

As emphasized earlier, France and Spain are among the strongest supporters of the CAP of the EU and have been over recent decades. As the data in Table 2.6 show, besides these two countries, Germany and Italy also account for an impressive share of the agricultural output value – more than 12% each, based on 2015 Eurostat data. This may come as no surprise, because, as described earlier, these countries benefit from large agricultural territories rather than large-farm holdings and have a considerable number of full-time workers in the field of agriculture. The Netherlands, which is smaller than Poland or Romania and accounts for fewer agricultural holdings, has a share of

Table 2.5. Agricultural income per AWU (year 2010 = 100). (From Eurostat, 2017b.)

Country/region	2010	2011	2012	2013	2014	2015	2016
EU-28	100.0	108.4	107.1	111.2	112.8	108.8	109.3
EU-27	100.0	108.5	107.3	111.3	113.0	108.8	109.2
Austria	100.0	115.1	108.0	94.7	87.9	81.6	92.6
Belgium	100.0	89.2	108.0	87.7	83.4	87.3	80.5
Bulgaria	100.0	114.1	133.3	162.0	172.8	158.1	188.2
Croatia	100.0	95.5	81.6	90.3	78.2	105.8	117.7
Cyprus	100.0	74.7	103.0	102.2	94.7	123.3	125.8
Czech Republic	100.0	134.8	133.6	134.9	155.3	137.7	155.1
Denmark	100.0	112.4	153.5	106.7	119.7	77.5	60.3
Estonia	100.0	124.5	144.8	134.6	127.4	103.2	65.2
Finland	100.0	86.4	87.6	86.3	77.9	64.8	69.5
France	100.0	104.6	105.3	89.6	101.8	108.6	95.6
Germany	100.0	112.2	101.0	115.6	109.4	72.4	81.3
Greece	100.0	87.2	87.1	80.6	88.9	97.1	92.4
Hungary	100.0	148.6	136.5	150.0	160.0	152.6	163.3
Iceland	100.0	109.6	121.0	98.3	156.4	175.7	168.1
Ireland	100.0	130.2	117.3	121.6	125.9	119.4	123.7
Italy	100.0	118.3	126.9	150.2	136.3	133.7	129.9
Latvia	100.0	95.8	115.2	103.9	115.7	131.2	120.3
Lithuania	100.0	127.8	159.6	144.6	131.9	145.2	120.5
Luxembourg	100.0	99.9	105.8	91.6	121.4	95.2	103.6
Malta	100.0	87.0	82.2	81.0	79.7	96.2	69.4
Netherlands	100.0	87.0	94.0	105.8	101.3	99.8	101.0
Norway	100.0	98.4	102.6	99.2	108.1	123.3	126.1
Poland	100.0	113.8	106.3	114.9	103.0	99.3	125.2
Portugal	100.0	83.6	91.8	106.9	108.2	110.7	130.6
Romania	100.0	127.9	94.4	111.5	121.8	120.7	118.1
Slovakia	100.0	118.7	133.7	130.3	143.3	142.7	173.2
Slovenia	100.0	113.9	92.0	92.8	105.3	115.7	104.9
Spain	100.0	101.1	102.5	112.7	116.3	119.7	123.0
Sweden	100.0	102.0	101.4	91.8	101.1	106.6	97.7
Switzerland	100.0	103.5	102.8	110.0	117.2	110.3	118.0
UK	100.0	116.5	110.4	119.2	116.2	98.2	95.4

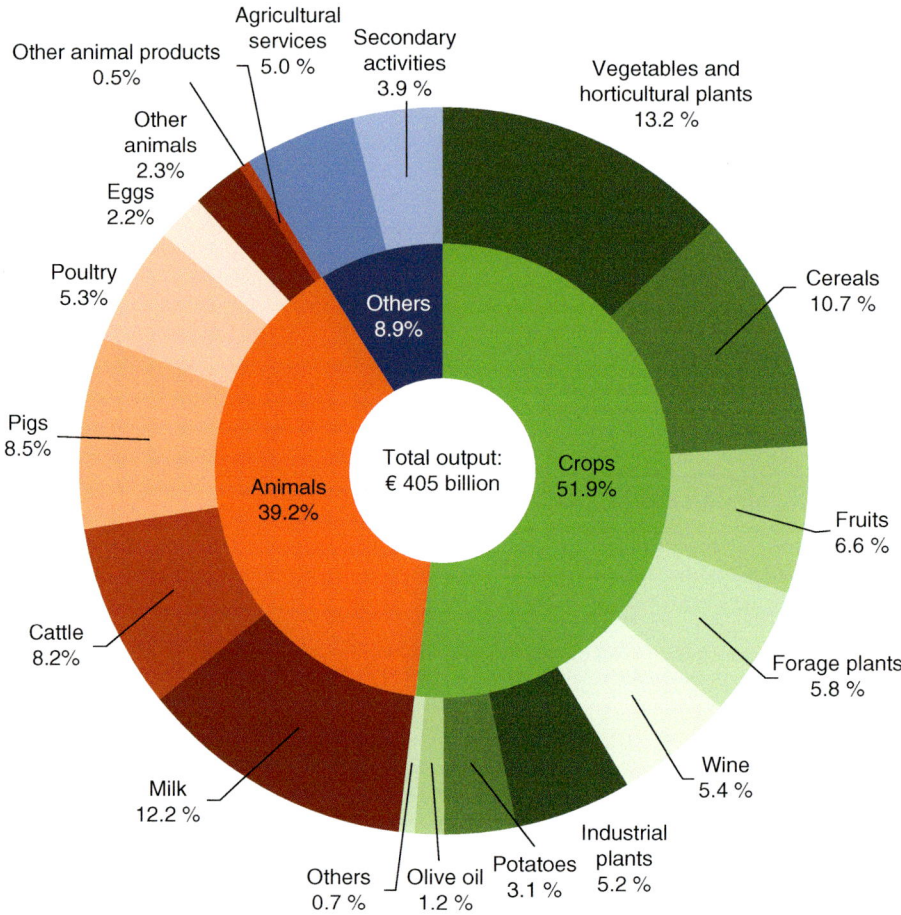

Fig. 2.11. Output of the agricultural industry, EU-28, 2016 (values at basic prices). (From Eurostat, 2016a.)

6.5% of the total agricultural output value; this proportion is almost double that of Romania, which accounted for more than one-third of the agricultural holdings of the EU, proving once more that intelligent agriculture is more effective than extensive agriculture.

Notes

[1] The six countries that founded the European Coal and Steel Community (treaty signed in 1951 in Paris and entered into force in 1952), the European Economic Community and the European Atomic Energy Community (treaties signed in 1957 in Rome and entered into force in 1958) were: Belgium, France, Italy, Luxembourg, The Netherlands and Germany (West Germany).
[2] The Treaty of Rome establishing the European Economic Community has been renamed by the Treaty of Lisbon (signed in 2007 in Lisbon and entered into force in 2009) as the Treaty on the Functioning of the European Union.

Table 2.6. Output value of the agricultural industry (values at basic prices), 2010 and 2015. (From Eurostat, 2016a.)

	Value (€ million)		Share of EU-28 (%)	
	2010	2015	2010	2015
EU-28	367,754.3	411,156.9	100.0	100.0
Austria	6,315.5	6,778.8	1.7	1.6
Belgium	7,758.2	8,116.8	2.1	2.0
Bulgaria	3,821.9	4,033.2	1.0	1.0
Croatia	2,914.3	2,277.4	0.8	0.6
Cyprus	685.7	693.4	0.2	0.2
Czech Republic	4,058.1	4,550.4	1.1	1.1
Denmark	9,740.9	10,269.2	2.6	2.5
Estonia	668.3	935.1	0.2	0.2
Finland	4,214.0	4,270.2	1.1	1.0
France	68,125.2	75,167.4	18.5	18.3
Germany	46,019.0	51,548.2	12.5	12.5
Greece	10,567.5	10,665.3	2.9	2.6
Hungary	6,121.8	7,925.5	1.7	1.9
Iceland	292.2	425.3	0.1	0.1
Ireland	5,822.0	7,397.1	1.6	1.8
Italy	48,159.8	55,203.9	13.1	13.4
Latvia	941.6	1,402.1	0.3	0.3
Lithuania	2,042.5	2,971.8	0.6	0.7
Luxembourg	325.3	401.4	0.1	0.1
Malta	126.1	128.1	0.0	0.0
Netherlands	25,318.7	26,708.2	6.9	6.5
Norway	4,626.4	5,507.4	1.3	1.3
Poland	19,768.8	22,320.2	5.4	5.4
Portugal	6,451.7	7,079.9	1.8	1.7
Romania	15,301.4	15,535.9	4.2	3.8
Slovakia	1,886.6	2,160.7	0.5	0.5
Slovenia	1,103.6	1,263.6	0.3	0.3
Spain	40,371.2	45,490.7	11.0	11.1
Sweden	5,379.0	6,239.5	1.5	1.5
Switzerland	7,278.9	8,323.0	–	–
UK	23,745.7	29,623.1	6.5	7.2

[3] On 23 June 2016, the population of the UK was asked to vote whether this country should continue as an EU Member State or leave the EU, and the result was in favour of exiting the EU.

[4] Although the UK has expressed the will to leave the EU and negotiations to do so based on Article 50 of the Treaty of Lisbon have begun, this country is still considered part of the EU until the leaving treaty is signed.

[5] ESU is a standard gross margin of €1200 that is used to express the economic size of an agricultural holding or farm: 1 ESU roughly corresponds to either 1.3 ha of cereals, one dairy cow or 25 ewes, or equivalent combinations of these (Thurston, 2008).

[6] One annual work unit (AWU) corresponds to the work performed by one person who is occupied on an agricultural holding on a full-time basis (Eurostat, 2015a).

3 Analysis of the European Agricultural Context[1]

3.1 Consideration of Romanian Agriculture within CAP Coordinates

Achieving a functional agriculture, based on the demands of the market economy and in line with the European model of agriculture and food production, requires the realization and implementation of a strong and sustainably financed national agricultural policy in the case of Romania that contributes directly to ensuring competitiveness in the agricultural sector, as well as guaranteeing a comparative standard of living for the large rural population and reducing regional inequities.

Agriculture and the agri-food sector represent for Romania and its national economy an important economic branch with significant implications and determinations in the evolution and realization of generally sustainable economic growth and poverty reduction, and equally is an essential component in the development of rural communities, serving to significantly reduce the regional and social gaps.

Agriculture contributes decisively in terms not only of strengthening the vitality of the rural environment and the rural communities involved but also of the stability of the economy as a whole, fulfilling decisively the exigencies imposed by the need for food security and safety.

As has been shown in many studies, agriculture represents an economic sector with a large impact on economic development in the Romanian economy, requiring a wide process of adaptation to the realities of the functional competitive market, which must simultaneously meet the need for harmonious development of the rural communities (Postoiu and Buşega, 2015; Tudor, 2015; Lazăr and Lazăr, 2016; Constantin, 2017; Feher et al., 2017; Croitoru et al., 2018).

Over the years, Romanian agriculture and the agri-food sector have undergone much transformation and reform to bring them closer to the demands of the European agricultural model. As emphasized by Cartwright (2017) and Anghelache (2018), after the reforms started in 1989, Romanian agriculture moved from well-organized and structured forms of functioning to a reorientation of land and agricultural property towards the private dimension, which proved to be negative and potentially destructive for this economic sector. Practically, the effects of the reform process on Romanian agriculture are still felt today, with individual farms and subsistence farms accounting for the largest share of the holding system, as discussed in Chapter 2 (this volume).

Although it has demonstrated a high degree of resilience and adaptation to these reforms, the Romanian agricultural sector still faces specific problems caused by the existence of a predominantly agrarian economy in the rural environment and agricultural overpopulation. In addition, the national agricultural sector, in terms of its agricultural multi-functionality, is significant for maintaining the diversity of habitats and landscapes, as well as for protecting the environment. As the literature highlights, 'agriculture is an important sector for Romania, given the large agricultural area owned, which can contribute significantly to the qualitative transformation of the Romanian economy' (Aceleanu *et al.*, 2015).

Considering the importance of the national agricultural sector in achieving the multiple economic demands, starting with ensuring food safety and security of the population and food for livestock, as well as continuing the determinant role played in the development of rural communities and the rural population's dependence on agricultural activities, an inclusive analysis of the role, impact, trends and perspectives of agriculture and the agrarian economy in the evolution of the national economy is of high importance.

The integration of Romania into the EU and its adaptation to the new demands imposed by the convergence between national and community policies has directly and immediately reflected on the national agricultural sector, which played a significant role in the construction of the national economy. Thus, Romanian agriculture has had to go through a complex and difficult process of reform and adaptation to the new conditions as part of the essential transformations needed in order for the national agriculture to meet the challenges of the European model of agriculture and food production.

Although some studies have shown that Romanian agriculture benefits from these necessary conditions and has all the features to become a unique instrument in achieving economic development (World Bank, 2008; Andrei *et al.*, 2015; Ciampi Stančová and Cavicchi, 2019), delays are still found in capitalizing on the national agricultural potential and its convergence with the European agricultural model.

Figure 3.1 shows the importance of agriculture in rural areas in Romania in relation to four economic dimensions: the population, employment, gross value added (GVA) and territory. The figure highlights the importance of agriculture in rural areas: in predominantly rural areas with a strong agricultural character, agriculture produces 36.65% of the GVA and employs more than half of the population in these areas (55.04%), accounting for 53.77% of the total population, with a share of 67.82% of the territory. Beyond these figures, we can state that the Romanian economy still has a strong agricultural component, with a

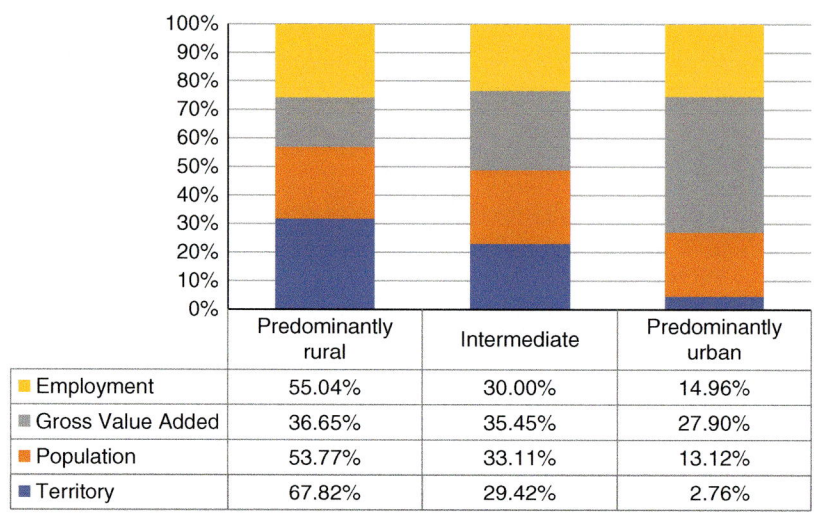

	Predominantly rural	Intermediate	Predominantly urban
▪ Employment	55.04%	30.00%	14.96%
▪ Gross Value Added	36.65%	35.45%	27.90%
▪ Population	53.77%	33.11%	13.12%
▪ Territory	67.82%	29.42%	2.76%

Fig. 3.1. Importance of rural areas and agriculture in Romania, 2017. (Based on information from European Commission, 2018b.)

mainly rural population, which should impose a broad reconsideration when addressing economic policies.

As noted by Andrei and Duşmănescu (2012), the main method of capitalizing on the national agricultural potential is through agricultural farms, as these are the only ones able to mobilize the labour force and the capital for exploitation of agricultural land. Figure 3.2 shows the distribution of farms and farmland by farm size in terms of standard output as a share of national totals. This shows the concentration of agricultural holdings at the two extremes of utilized agricultural area (UAA[2]) – very small farms, of which there is the highest number and which use 37.9% of the UAA, and large farms, which, despite their small share of only 0.1%, exploit 29.8% of the UAA. As Bularca (Olaru) and Toma (2018) found, Romanian agriculture is dominated by very small family farms that, despite limited economic viability, are a source of sustainability in the national economy as a whole, providing broad support for the viability of rural communities, rural and local culture, and environmental sustainability of agriculture, as well as sources of income and survival for the rural population.

In this context, the transformations of the national agricultural sector have been determined partly by changes in the CAP paradigm but mainly by the manifestation of broader influences and complex phenomena, such as the integration and globalization of the markets under the pressure derived from changes in the global economy. From this perspective, analysis of the transformations of the national and European agricultural sector, of the agricultural market and of the European agricultural model under the influence of the reform of the CAP is required.

Figure 3.3 and Table 3.1 show the efficiency of absorption of financial allocations within the CAP and its level of achievement in the case of Romanian

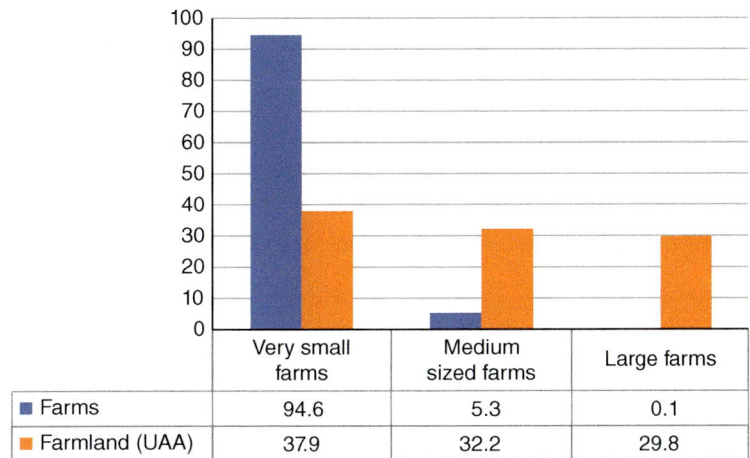

	Very small farms	Medium sized farms	Large farms
■ Farms	94.6	5.3	0.1
■ Farmland (UAA)	37.9	32.2	29.8

Fig. 3.2. Distribution of farms and farmland (utilized agricultural area (UAA)) in Romania by farm size in terms of standard output (% share of national totals), 2016. (Based on information from European Commission, 2018b.)

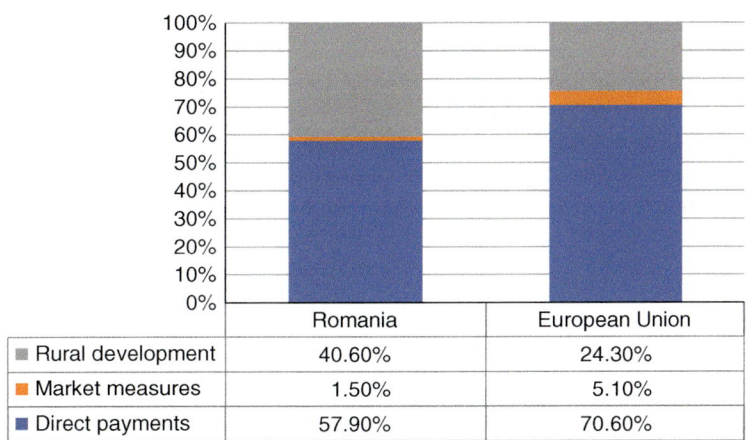

	Romania	European Union
▤ Rural development	40.60%	24.30%
■ Market measures	1.50%	5.10%
■ Direct payments	57.90%	70.60%

Fig. 3.3. Distribution of CAP expenditure in Romania versus the EU, 2017. (Based on information from European Commission, 2018b.)

agriculture. Financial allocations made through the CAP represent significant support for improving the conditions in Romanian agriculture. How governments respond to CAP flexibility provided by its most recent reforms will also determine the reorientation of available rural resources. Henke *et al.* (2018) noted significant changes and reorientations in addressability of financial measures under the CAP. In this context, in Fig. 3.3, it can be seen that there are significant differences in the three indicators in the case of Romania compared with the EU. Although rural development managed to achieve a higher level than that of the EU, in the case of direct payments and market measures the

Table 3.1. Distribution of CAP expenditure in Romania according to the destination of funds, 2017. (Based on information from European Commission, 2018b.)

Measure	Absorbed	Allocated	Difference
Decoupled direct aid	42.60%	73.50%	30.90%
Other direct aid	15.30%	26.50%	11.20%
Additional amounts of aid	0.00%	0.00%	0.00%
Reimbursement of direct aid in relation to financial discipline	0.00%	0.00%	0.00%
Direct payments	**57.90%**	**100.00%**	**42.10%**
Fruit and vegetables	0.20%	15.20%	15.00%
Wine sector	0.40%	26.90%	26.50%
Promotion	0.00%	0.60%	0.60%
Milk and milk products	0.50%	36.30%	35.80%
Pork, eggs, poultry and others	0.30%	21.00%	20.70%
Market measures	**1.50%**	**100.00%**	**98.50%**
Rural development	**40.60%**	**100.00%**	**59.40%**

situation was clearly unfavourable to Romania, showing modest levels below the European level. To better understand this situation, Table 3.1 shows the distribution of CAP expenditure in Romania in terms of funding destinations in 2017, with the three measures of Fig. 3.3 shown in bold.

The distribution of CAP expenditure in Romania according to destination in 2017 shows that there was poor absorption of funds for agriculture, contributing to widening of the gaps between the national agricultural sector and the European agricultural model. This low efficiency of absorption of CAP funding weakens the national agricultural sector, depriving it from the financing needed to increase the level of convergence with European requirements in the field and limiting its competitiveness in practical terms.

The convergence potential was analysed by Ciutacu *et al.* (2015) in terms of the significant differences and existing disparities between the national and the European agricultural sectors. Thus, the main objectives of the study were to highlight the differences and similarities between the agricultural model and the European rural development model, while analysing the current situation in the Romanian agricultural sector. By taking into consideration the agricultural statistical indicators for the last two decades, the authors placed Romania outside the frame of the economic and agricultural space of the EU Member States, with minor similarities but very strong discrepancies between the economic, technical and institutional characteristics.

One of the conclusions of this study by Ciutacu *et al.* (2015) reiterated the fact that ensuring the development of the agricultural sector in Romania and eliminating the existing gaps compared with the European agricultural model cannot be achieved by miracles. Obtaining a high level of economic convergence and symmetry requires a set of policies that will address technical, technological, economic, institutional, cultural, educational and social issues, all within a synergistic approach.

3.2 Methodological Specifications

To analyse and highlight the evolution of the agricultural sector and the European agricultural industry in the case of several EU-28 Member States and Norway, a time frame between 2006 and 2015 was chosen and 96 data series were selected; 92 data series referred to 23 EU-28 Member States with available data for the following four indicators: vegetal production (PV), animal production (PA), total labour force (TLF) and GVA in the agri-food sector (VABA, i.e. value added by agriculture), while four data series corresponded to information for Norway.

The objectives of the research were, on the one hand, to determine and investigate the clusters formed by considering the four indicators analysed for the years 2006, 2009 and 2015, and, on the other hand, to analyse the similarities and differences between the states that formed representative clusters for the conducted research. For each of the three years for which the cluster analysis was performed, a matrix, X, was formed (Greene, 2000; Andren, 2007; Dougherty, 2007; Zaharia and Popescu, 2015), structured as follows:

$$X = \langle x_{ij} \rangle_{i=\overline{1,n}, j=\overline{1,m}} \tag{1}$$

where $n = 24$, representing the number of countries, and $m = 4$, representing the number of indicators (variables) taken into consideration. For each element of the matrix X, a Z-score transformation was applied:

$$y_{ij} = \frac{x_{ij} - \overline{x}_j}{S_j} \tag{2}$$

where

$$\overline{x}_j = \frac{\sum_{i=1}^{n} x_{ij}}{n}$$

and

$$S_j = \sqrt{\frac{\sum_{i=1}^{n} (x_{ij} - \overline{x}_j)^2}{n-1}}.$$

Cluster generation was performed using hierarchical clustering (Ketchen and Shook, 1996; Saint-Arnaud and Bernard, 2003; Rotariu *et al.*, 2006; Sánchez-Garcia *et al.*, 2014). Among the available methods for determining the distance between the elements, the best results were obtained using the Chebyshev method (Gutiérrez and Hernández, 1997; Peyret, 2002), while for generating the clusters, the furthest-neighbour method was used (Pagh *et al.*, 2015).

For comparability purposes, in addition to hierarchical clustering, for 2006, the *K*-means cluster methodology was also employed (Aggarwal and Reddy, 2013; Ghosh and Dubey, 2013; Dipti and Patel, 2014).

Analysis of variance (ANOVA) was used to test the statistical significance of how each variable belongs to a cluster (Cardinal and Aitken, 2013; Gu, 2013; Lawal, 2014; Walde, 2014). The hypotheses of the test were:

- H_{0_1}: the analysed variable is not significant in relation to cluster membership (the variance of the mean square between groups (MSB) does not differ significantly from the variance of the mean square within groups (MSW)).
- H_{1_1}: the analysed variable is significant in relation to cluster membership (the variance between groups is significantly higher than the variance within groups).

The null hypothesis (H_{0_1}) is accepted if the statistics of the *F*-test fulfil the following relationship (Cornish, 2006; Zaharia and Popescu, 2015):

$$F = \frac{MSB}{MSW} < F_{\infty, r-1, n-r-1}, \text{ equivalent to Sig. } F > \alpha \tag{3}$$

In Eqn (3):

$$MSB = \frac{\sum_{i=1}^{r} (\bar{y}_i - \bar{y})^2 \times n_i}{r-1}$$

and

$$MSW = \frac{\sum_{i=1}^{r} \sum_{j=1}^{n_i} (y_{ij} - \bar{y}_i)}{n-r-1}$$

where *r* represents the number of groups and n_i is the number of elements in each group.

One of the most important parameters of the data series is the mean of the series. To be considered in characterizing the evolution of the four variables for each state, its statistical significance was tested according to the following hypotheses:

- H_{0_2}: the mean value of the analysed variable is not statistically significant.
- H_{1_2}: the mean value of the analysed variable is statistically significant.

Depending on the number of elements of the data series (*n*) and on the significance threshold (α), the limits of the confidence interval (CI) of the mean are: lower bound (LB) and upper bound (UB), given by the relationship (Zaharia and Gogonea, 2009; Kerenidis *et al.*, 2012; Araya *et al.*, 2014):

$$LB = \bar{x} - \frac{S}{\sqrt{n}} \cdot t_{\frac{\alpha}{2}, n-1} \tag{4a}$$

and

$$UB = \bar{x} + \frac{S}{\sqrt{n}} \cdot t_{\frac{\alpha}{2}, n-1} \tag{4b}$$

where

$$s = \sqrt{\frac{\sum_{i=1}^{n} (x_i - \bar{x})^2}{n-1}}$$

The null hypothesis (H_{0_2}) is accepted if the statistics of the *t*-test fulfils the relationship:

$$t_{st} \in \left(-t_{\frac{\alpha}{2}, n-1}, t_{\frac{\alpha}{2}, n-1} \right) \tag{5}$$

where

$$t_{st} = \frac{\overline{x}}{\frac{s}{\sqrt{n}}}$$

Given that Eqn (5) is fulfilled, then the following relation is also satisfied:

$$\text{Sign}(LB) \neq sign(UB) \tag{6}$$

For clusters with more than two elements, the statistical significance of the mean values of the analysed variables determined for each cluster was also tested. In this case, the tested hypotheses are:

- H_{0_3}: the mean value of the analysed variable is not statistically significant in relation to cluster membership.
- H_{1_3}: the mean value of the analysed variable is statistically significant in relation to cluster membership.

In this research, the null hypothesis H_{0_3} was accepted if the condition of Eqn (6) was fulfilled, namely if the limits of the CI of that variable's mean determined for each cluster were of different signs. Moreover, for each country included in the significant clusters, the Pearson correlation coefficient was determined, and their statistical significance was analysed using the *t_bilateral* test. The hypotheses of the test were:

- H_{0_4}: the value of the correlation coefficient is not statistically significant.
- H_{1_4}: the value of the correlation coefficient is statistically significant.

The requisite for accepting the null hypothesis H_{0_4}, for a significance threshold α, is:

$$Sig.(2\_\text{tailed}) > \alpha \tag{7}$$

For each country within the clusters considered significant, potential functional links between the GVA in the agri-food sector (VABA; i.e. value added by agriculture) and the other three variables (PA, PV and TLF) were tested using models such as the dependence evolution model of VABA (MVABAx) (Zaharia and Gogonea, 2009; Zaharia and Popescu, 2015):

$$\text{MVABAx} = c_0 + \sum_{i=1}^{k} c_i \times x_i + \varepsilon, \ c_0, \ c_i \in R \tag{8}$$

In this situation, the tested hypotheses were:

- H_{0_5}: model (8) is not statistically valid.
- H_{1_5}: model (8) is statistically significant.

Thus, the condition for accepting the null hypothesis H_{0_5}, for a significance threshold α, becomes:

$$F = \frac{\text{MSR}}{\text{MSE}} < F_{\infty,k,n-k-1} \text{ equivalent to Sig. } F > \alpha \tag{9}$$

where MSR is the regression mean square and MSE is the mean square error.

The terms defined in relation to Eqn (9) are accordingly:

$$\text{MSR} = \frac{\sum_{i=1}^{n}(\hat{y}_i - \bar{y})^2}{k}, \ \text{MSE} = \frac{\sum_{i=1}^{n}(y_i - \hat{y}_i)^2}{n-k-1}, \ \hat{y}_i = c_0 + \sum_{i=1}^{k} c_i \times x_i$$

Finally, in order to test the statistical significance of the values of the parameters in model (8), the tested hypotheses were:

- H_{0_6}: the value of the parameter c_i in model (8) is not statistically significant.
- H_{1_6}: the value of the parameter c_i in model (8) is statistically significant.

The null hypothesis H_{0_6} was considered to be accepted if the condition of Eqn (6) was satisfied for the limits of the CI of the respective parameter.

Data processing was carried out using the automated data processing software SPSS 14 (Howitt and Cramer, 2005; Popa, 2008) and the Excel Data Analysis module from the MS Office pack (Zaharia and Oprea, 2011).

To achieve cluster analysis of production, labour and value added in the agri-food sector, grouping of the 24 analysed countries was built on the performance level of the agri-food industry as well as on the production outputs in the years 2006, 2009 and 2015 according to four indicators:

- PA (animal production): measured as production prices in million €;
- PV (vegetal production): measured as production prices in million €;
- TLF (total labour force input in the agri-food sector): measured as absolute values in 1000 AWUs;
- VABA (gross value added in the agri-food sector): measured as producer prices in million €.

3.3 Cluster Analysis and Grouping of EU Countries in 2006

The first significant clusters were attained by first generating and grouping the analysed states into three clusters. In this situation, the first cluster (A1) consists of seven countries (Belgium, Czech Republic, Ireland, Austria, Finland, Sweden and Norway), the second (A2) of six countries (Estonia, Latvia, Lithuania, Luxembourg, Slovenia and Slovakia) and the third (A3) of two countries (Hungary and Portugal).

For the chosen significance level (95% CI), by considering the results of the variance analysis (Table 3.2), the null hypothesis H_{0_1} was rejected and the alternative hypothesis H_{1_1} was accepted. Consequently, all four variables were significant in terms of cluster membership. This fact was highlighted on the one hand by the values of the F statistics, which in all four cases presented higher values than the critical value $F_{0.05;\ 41;\ 43} \cong 1.508$, but also, on the other hand, by the fact that Sig. $F = 0.000 < \alpha = 0.05$, which consequently leads to the rejection of the null hypothesis and acceptance of the alternative hypothesis.

The mean, minimum and maximum values of the four variables for the clusters, as well as the 95% CIs of the mean values for the chosen significance level, are presented in Table 3.3. From a graphical point of view, the main features of clusters A1–A3 are illustrated in Fig. 3.4.

Table 3.2. ANOVA table for testing the statistical significance of PA, PV, TLF and VABA corresponding to 12 cluster groupings in 2006. df, degrees of freedom; Sig, significance.

Variable		Sum of squares	df	Mean square	*F*	Sig.
PA	Between groups	900,959,147.266	11	81,905,377.024	181.246	0.000
	Within groups	5,422,807.895	12	451,900.658		
	Total	906,381,955.162	23			
PV	Between groups	1,970,669,411.441	11	179,151,764.676	671.608	0.000
	Within groups	3,201,004.534	12	266,750.378		
	Total	1,973,870,415.975	23			
TLF	Between groups	10,610,087.599	11	964,553.418	324.554	0.000
	Within groups	35,663.256	12	2,971.938		
	Total	10,645,750.855	23			
VABA	Between groups	1,312,636,073.799	11	119,330,552.164	644.448	0.000
	Within groups	2,222,006.023	12	185,167.169		
	Total	1,314,858,079.822	23			

A first observation that emerged from analysis of the data in Table 3.3 was that, for the PA and PV variables, all the mean values in each of the three clusters and the values of the TLF and VABA variables for clusters A1 and A2 generate a rejection of the null hypothesis (H_{0_3}) and acceptance of the alternative hypothesis (H_{1_3}), the mean values determined at the respective clusters being statistically significant. This conclusion stems from the fact that the values of the minimum and maximum limits of the CIs of each variable have the same sign.

Cluster A1 was characterized by high mean values of PA and PV, resulting in significant values of VABA under the conditions of low values of TLF.

The mean value of PA in the seven countries of cluster A1 was €2461.50 million, with a 95% CIs of €1611.39 million to €3311.61 million.

The value of 37.34% for the coefficient of variation, although higher than 30%, reflected a relatively good representability of the mean, being influenced by the significant differences between the lowest value of PA of €1592.97 million recorded in 2006 in the Czech Republic, and €1646.27 million in Finland, against the high values in Ireland (€3766.95 million) and Belgium (€3644.55 million).

Another characteristic of cluster A1 was that the mean value of PV was about 24.6% lower than the mean value of PA; the 95% CI for the mean value of PV was €1176.31 million to €2447.97 million.

The value of the coefficient of variation for PV (37.94) did not differ significantly from that corresponding to PA, the representativeness of the mean value also being quite appropriate in this case. The lowest value of PV for the seven states of cluster A1 was recorded in Finland (€1122.1 million), while the highest value (€3163.83 million) was determined in Belgium.

The mean value of VABA in cluster A1 was €1362.21 million with a 95% CI of €832.03 million to €1892.39 million and a coefficient of variation of 42.08%, the lowest value of VABA being recorded in Finland (€736.42 million) and the highest value in Belgium (€2145.75 million).

Table 3.3. Mean, minimum and maximum values, 95% confidence intervals (CIs) and variation coefficients of PA, PV, TLF and VABA for clusters A1–A3 in 2006. Note that clusters A1–A3 contain a total of 15 states. At this point in the cluster generation, the other nine states are not included in the clusters. SD, standard deviation; SE, standard error.

Variable	Cluster	N	Mean	SD	SE	95% CI for mean		Min.	Max.
						LB	UB		
PA	A1	7	2461.50	919.19	347.42	1611.39	3311.61	1592.97	3766.95
	A2	6	451.08	252.38	103.03	186.22	715.94	144.16	774.73
	A3	2	2133.29	186.71	132.03	455.75	3810.82	2001.26	2265.31
PV	A1	7	1812.14	687.50	259.85	1176.31	2447.97	1122.10	3163.83
	A2	6	455.58	265.90	108.55	176.53	734.63	124.52	792.94
	A3	2	3408.97	107.38	75.93	2444.19	4373.75	3333.04	3484.90
TLF	A1	7	103.45	37.43	14.15	68.83	138.07	63.30	152.90
	A2	6	84.97	58.07	23.71	24.02	145.91	3.86	165.80
	A3	2	432.31	101.95	72.09	-483.68	1348.30	360.22	504.40
VABA	A1	7	1362.21	573.26	216.67	832.03	1892.39	736.42	2145.75
	A2	6	284.09	130.05	53.09	147.61	420.57	96.90	409.50
	A3	2	2088.15	407.01	287.80	-1568.70	5745.00	1800.35	2375.95

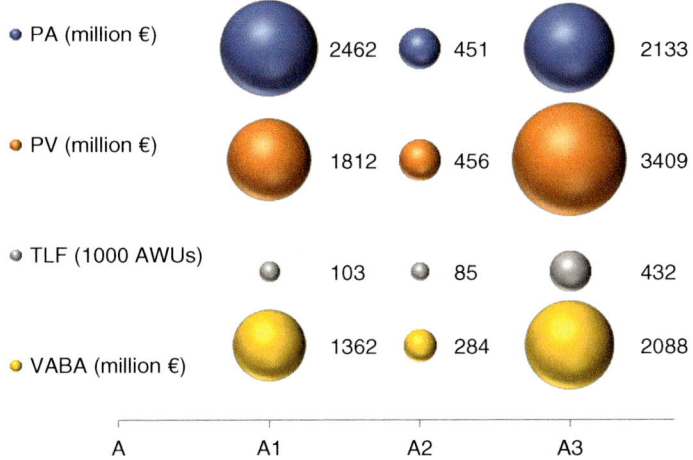

Fig. 3.4. Comparative presentation of the values of the PA, PV, TLF and VABA variables in clusters A1, A2 and A3 in 2006. (Based on information from Eurostat, 2017a,b.)

Regarding TLF, expressed in absolute terms (1000 AWUs), in cluster A1 the mean value was 103.45; the lowest value for this indicator was registered in Norway (63.3) and the highest in Ireland (152.9).

Ultimately, if we express labour productivity in agriculture (LPA) as a ratio between VABA and TLF, then the highest value of LPA in cluster A1 was registered in Belgium (€31.56 million/1000 AWUs) and the lowest in the Czech Republic (€7.28 million/1000 AWUs) and Finland (€7.91 million/1000 AWUs).

Cluster A2, consisting of six states (Estonia, Latvia, Lithuania, Luxembourg, Slovenia and Slovakia), was characterized by mean values of the four analysed variables that were considerably lower than those in cluster A1. This aspect is a consequence of the smaller areas of these countries and their characteristics.

Another characteristic of cluster A2 was that the individual values of the analysed variables recorded in 2006 in its countries had a much larger scatter than in cluster A1. This led to values of coefficients of variation much higher than 30% (55.95% for PA, 58.37% for PV, 45.78% for VABA and 68.45% for TLF), which determined a reduced representability of the means recorded by these variables.

The mean value of PA in the six countries of cluster A2 was €451.08 million, with a 95% CI of €186.22 million to €715.94 million. The lowest value of PA in this cluster was €144.16 million, seen in 2006 in Luxembourg, and the maximum was €774.73 million in Slovakia.

A third characteristic of cluster A2 was that, unlike cluster A1, the mean value of PV (€455.58 million) was sensibly equal to that of PA. For the mean value of PV, the 95% CI was €176.53 million to €734.63 million. The lowest and highest values were recorded in Luxembourg (€124.52 million) and Slovakia (€792.94 million), respectively.

The mean value of VABA in cluster A2 was €284.09 million with a 95% CI of €147.61 million to €420.57 million, with the lowest value being recorded in Luxembourg (€95.90 million) and the highest value in Lithuania (€409.5 million).

Regarding TLF, the mean value for the A2 cluster was 84.97, which was also lower compared with the corresponding mean value in cluster A1. In 2006, the lowest value of this indicator was confirmed in Luxembourg (3.86) and the highest in Lithuania (165.86).

In terms of LPA, values ranged between €1.81 million/1000 AWUs in Latvia and €25.1 million/1000 AWUs in Luxembourg. In conclusion, in cluster A2, in terms of the values of PA and PV, Luxembourg ranked last. However, in 2006, Luxembourg recorded a productivity that was six times higher than that in Slovakia, which accounted for the highest values for both PA and PV in the same year in cluster A2.

Cluster A3 consisted of only two states: Hungary and Portugal. In this cluster, both PA and PV values were relatively close. In Hungary, in 2006, PA was €2001.26 million and PV was €3333.04 million; in the same year, in Portugal, PA was €2265.31 million and PV was €3484.9 million.

There were significant differences between these two countries in terms of VABA, whose value was €2375.95 million in Portugal compared with only €1800.35 million in Hungary. These differences also characterized the ratio between VABA and TLF, with LPA values of €360.22 million/1000 AWUs in Portugal and €504.40 million/1000 AWU in Hungary. Consequently, although there were no significant differences in PA and PVC between the two countries, LPA was significantly different within the two countries in cluster A3. In Portugal, in 2006, LPA was €6.6 million/1000 AWUs, while in Hungary it was almost half this value (€3.57 million/1000 AWUs).

Continuing the grouping process using the hierarchical cluster method, the new clustering comprised C1–C4. For C1, Bulgaria was included in cluster A2, which was joined with the A1 and A3 clusters and to which the Netherlands, the UK and Germany were also added, so that this cluster contained 19 states. In order to group the 24 states into four clusters (the minimum limit chosen), the remaining clusters contained two (C2), one (C3) and two (C4) states. The main parameters of the four clusters are shown in Table 3.4.

Cluster C1 was formed by the union of clusters A1, A2 and A3 whose characteristics were analysed previously, as well as Bulgaria, the Netherlands, the UK and Germany. This led to extremely high values of the coefficient of variation, which determined the loss of representativeness of the mean of all four variables.

Cluster C2 consisted of two states, Spain and Italy, and was characterized by significantly high values for both PA and PV; in addition, high values of VABA (an average of €24,309.5 million) led to high productivity (LPA values of €20.08 million/1000 AWUs in Spain and €21.37 million/1000 AWUs in Italy). These features are also highlighted by the graphical representation of the parameter values analysed for clusters C1–C4 shown in Fig. 3.5.

Taking into account the values of the coefficient of variation, which were much lower than 30% (1.9% for PA, 15.28% for PV, 19.53% for TLF and 15.18%

Table 3.4. Mean, minimum and maximum values, 96% CIs and variation coefficients of PA, PV, TLF and VABA for clusters C1–C4 in 2006.

Variable	Cluster	N	Mean	SD	SE	95% CI for mean		Min.	Max.
						LB	UB		
PA	C1	19	3,406.8	4,869.9	1,117.2	1,059.5	5,754.0	144.2	19,728.8
	C2	2	13,513.2	256.1	181.1	11,212.0	15,814.4	13,332.1	13,694.3
	C3	1	21,696.2	–	–	–	–	21,696.2	21,696.2
	C4	2	5,897.8	2,649.4	1,873.4	−17,905.7	29,701.4	4,024.4	7,771.2
	Total	24	5,218.6	6,277.6	1,281.4	2,567.8	7,869.4	144.2	21,696.2
PV	C1	19	3,272.6	4,712.6	1,081.1	1,001.2	5,544.0	124.5	18,858.3
	C2	2	24,309.5	3,715.0	2,626.9	−9,068.2	57,687.3	21,682.6	26,936.4
	C3	1	33,361.7	–	–	–	–	33,361.7	33,361.7
	C4	2	8,363.4	737.9	521.8	1733.7	14,993.1	7,841.6	8,885.1
	Total	24	6,703.6	9,263.9	1,891.0	2,791.8	10,615.4	124.5	33,361.7
TLF	C1	19	194.0	177.3	40.7	108.5	279.5	3.9	568.0
	C2	2	1,135.1	172.3	121.9	−413.2	2,683.5	1,013.3	1,257.0
	C3	1	886.2	–	–	–	–	886.2	886.2
	C4	2	2,409.5	166.2	117.6	915.8	3,903.1	2,291.9	2,527.0
	Total	24	485.9	680.3	138.9	198.6	773.2	3.9	2,527.0
VABA	C1	19	2,370.3	3,355.6	769.8	753.0	3,987.7	96.9	12,919.7
	C2	2	23,604.4	4,609.9	3,259.7	−17,814.1	65,022.9	20,344.7	26,864.1
	C3	1	21,346.1	–	–	–	–	21,346.1	21,346.1
	C4	2	5,581.2	596.7	421.9	220.1	10,942.2	5,159.2	6,003.1
	Total	24	5,198.1	7,560.9	1,543.4	2,005.4	8,390.8	96.9	26,864.1

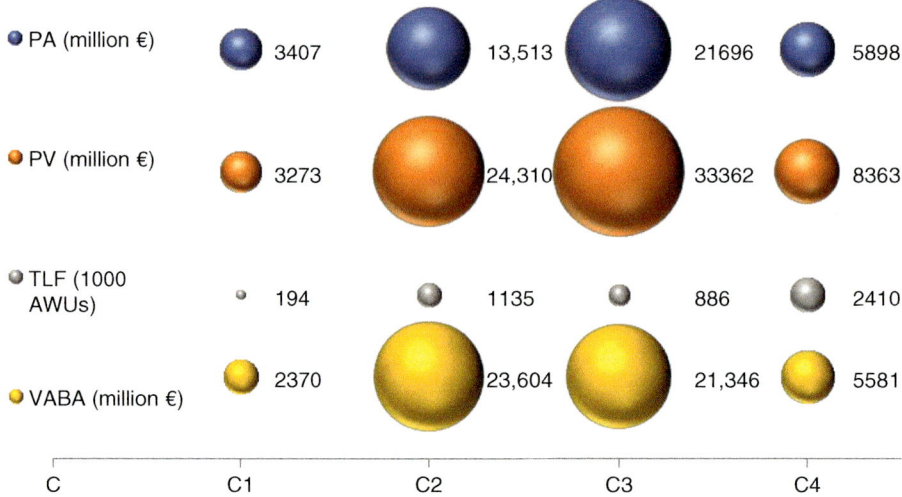

| | C | C1 | C2 | C3 | C4 |

Fig. 3.5. Comparative presentation of the values of the variables PA, PV, TLF and VABA in clusters C1–C4 in 2006. (Based on information from Eurostat, 2017a,b.)

for VABA), a viable conclusion was that the mean values of the four variables were very representative.

Continuing our study, cluster C3 contained only one state, France, which was characterized by the highest values of both PA and PV of all 24 countries analysed. In terms of LPA, France ranked second (€24.04 million/1000 AWUs), after the Netherlands.

Lastly, cluster C4 consisted of Romania and Poland and was characterized by high values of TLF of 2291.9 in Poland and 2527.0 in Romania, which, in 2006, led to some of the lowest productivities: only €2.25 million/1000 AWUs in Poland and €2.38 million/1000 AWUs in Romania.

A relatively similar group was obtained using the K-means cluster methodology (K1–K4); also in this case, for the significance level chosen previously (95% CI) and considering the results of the variance analysis (Table 3.5), all four variables were significant from the point of view of cluster membership.

The values of the F statistics also confirmed this assumption, which in all three cases presented higher values than the critical value of $F_{0.05;41;43} \cong 1.508$; moreover, Sig. $F = 0.000 < \alpha = 0.05$, for the variables PA, PV and VABA, while Sig. $F = 0.003 < \alpha = 0.05$ for the variable TLF.

The mean, minimum and maximum values of the four variables in clusters K1–K4, as well as the 95% CIs of the mean values are presented in Table 3.6, while a comparative arrangement of the mean values of the variables analysed in these clusters are represented graphically in Fig. 3.6.

One notable difference between the groups obtained using the hierarchical cluster methodology (Table 3.4) and the K-means cluster methodology (Table 3.6) was that the number of states included decreased from 19 in cluster C1 to 16 in cluster K1 by not including Germany, the Netherlands and the UK, resulting in a significant improvement of the standard deviation from 4869.9 to

Table 3.5. ANOVA table for testing the statistical significance of PA, PV, TLF and VABA corresponding to four cluster groupings using *K*-means cluster methodology for 2006.

Variable		Sum of squares	df	Mean square	F	Sig.
PA	Between groups	806,971,014.870	3	268,990,338.290	54.117	0.000
	Within groups	99,410,940.292	20	4,970,547.015		
	Total	906,381,955.162	23			
PV	Between groups	1,922,134,731.747	3	640,711,577.249	247.687	0.000
	Within groups	51,735,684.228	20	2,586,784.211		
	Total	1,973,870,415.975	23			
TLF	Between groups	5,279,187.839	3	1,759,729.280	6.558	0.003
	Within groups	5,366,563.016	20	268,328.151		
	Total	10,645,750.855	23			
VABA	Between groups	1,258,038,457.249	3	419,346,152.416	147.606	0.000
	Within groups	56,819,622.573	20	2,840,981.129		
	Total	1,314,858,079.822	23			

Table 3.6. Mean, minimum and maximum values and 95% CIs of PA, PV, TLF and VABA for clusters K1–K4, obtained by *K*-means cluster methodology for 2006.

Variable	Cluster	N	Mean	SD	SE	95% CI for mean LB	UB	Min.	Max.
PA	K1	16	1,582	1,136	284	977	2,187	144	3,767
	K2	2	16,530	4,523	3,198	−24,108	57,169	13,332	19,729
	K3	2	17,695	5,658	4,001	−33,142	68,532	13,694	21,696
	K4	4	7,871	3,033	1,516	3,045	12,697	4,024	11,426
	Total	24	5,219	6,278	1,281	2,568	7,869	144	21,696
PV	K1	16	1,500	1,089	272	919	2,080	125	3,485
	K2	2	20,270	1,997	1,412	2,327	38,214	18,858	21,683
	K3	2	30,149	4,543	3,213	−10,671	70,970	26,936	33,362
	K4	4	9,013	1,763	881	6,208	11,818	7,781	11,546
	Total	24	6,704	9,264	1,891	2,792	10,615	125	33,362
TLF	K1	16	166	164	41	79	254	4	564
	K2	2	791	315	223	−2,038	3,619	568	1,013
	K3	2	1,072	262	185	−1,284	3,427	886	1,257
	K4	4	1,319	1,264	632	−693	3,331	160	2,527
	Total	24	486	680	139	199	773	4	2,527
VABA	K1	16	1,060	767	192	651	1,469	97	2,376
	K2	2	16,632	5,250	3,712	−30,539	63,804	12,920	20,345
	K3	2	24,105	3,902	2,759	−10,951	59,162	21,346	26,864
	K4	4	6,580	1,318	659	4,483	8,676	5,159	8,235
	Total	24	5,198	7,561	1,543	2,005	8,391	97	26,864

1136. Under these circumstances, the mean value of PA decreased from €3406 million in C1 to €1582 million in K1 and the value of PV from €3272 million in C1 to €1500 million in K1.

In contrast, the differences between the lowest and highest values of the analysed variables varied widely. Thus, the value of PA varied between €144

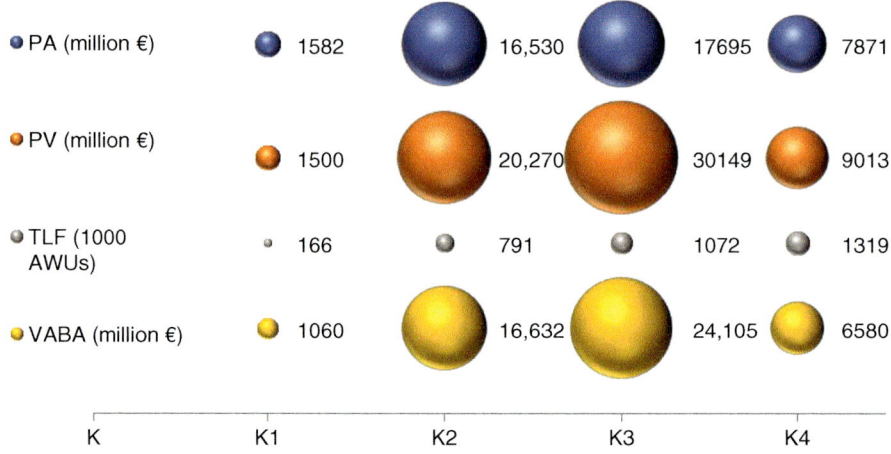

Fig. 3.6. Comparative presentation of the values of the variables PA, PV, TLF and VABA in clusters K1–K4 in 2006. (Based on information from Eurostat, 2017a,b.)

million in Luxembourg and €3767 million in Ireland, the value of PV varied between €125 million in Luxembourg and €3485 million in Portugal, TLF (in 1000 AWUs) varied between 4 in Luxembourg and 564 in Bulgaria, and VABA fluctuated between €97 million in Luxembourg and €2376 million in Portugal.

Cluster K2 consisted of Germany and Spain, and the 9.85% value of the coefficient of variation for PV was very representative of the mean of this indicator (€20,270 million).

In addition, given that for the other variables the coefficients of variation of the mean values were also approximately 30%, the conclusion could be drawn that, in this cluster, the mean values had a good representability. In cluster K2, the maximum value of PA (€19,729 million) was recorded in Germany, while the maximum values of PV (€21,683 million), TLF (1013, in 1000 AWUs) and VABA (€20,345 million) were recorded in Spain.

Cluster K3 consisted of France and Italy. In this cluster, given the values of the coefficient of variation (about 15%), the mean values of PV (€30,149 million) and VABA (€24,105 million) were very representative. In the K3 cluster, the maximum values of PA (€21,629 million) and PV (€33,362 million) were recorded in France, and of TLF (1257, in 1000 AWUs) and VABA (€26,864 million) were encountered in Italy.

Cluster K4 consisted of four countries, the Netherlands, Poland, Romania and the UK. In this cluster, except for the TLF variable for which the mean value was not representative due to the very large difference between the minimum value (in the Netherlands) and the maximum value (in Romania), the mean values of the other variables had a good representability.

In cluster K4, the maximum values of PA (€11,426 million) were recorded in the UK, of PV (€11,546 million) and of VABA (€8,235 million) were recorded in the Netherlands, and of TLF (2727, in 1000 AWUs) in Romania where LPA in 2006 was only €2.38 million/1000 AWUs, which is one of the smallest values among the countries analysed.

3.4   Cluster Analysis and the European Agricultural Model for 2009, a Year of Economic Crisis

In 2009, the agricultural sector in all 24 of the analysed countries was influenced by the economic crisis, with decreases in both PA and PV.

In the case of PA, the largest reduction in the value of production was in the Czech Republic, where PA decreased by 24%, from €1369.2 million in 2008 to €1125.3 million in 2009, while the smallest (almost insignificant) reduction was 0.5% in Romania, from €3851.9 million in 2008 to €3832.8 million in 2009.

For PV, the largest reduction in the value of production was in Romania where it decreased by 32.14%, from €12,421.2 million in 2008 to €8428.4 million in 2009, and the lowest reduction was 2.76% in Portugal, from €3447.7 million in 2008 to €3352.7 million in 2009.

Concerning TLF, with the exception of Romania and Hungary, where it remained approximately constant, there were reductions in the other 22 states ranging from 1.41% in the UK (from 287.38 in 2008 to 283.33 in 2009, in 1000 AWUs) to 9.09% in Norway (from 54.0 in 2008 to 51.4 in 2009).

These developments led to a 14.71% reduction in VABA in the EU-28, and among the 24 EU-28 countries under analysis, the highest reduction of 50.54% was recorded in Slovakia, from €460.05 million in 2008 to €227.52 million in 2009, while the lowest was 2.0% in Slovenia, from €391.93 million in 2008 to €383.95 million in 2009.

In order to determine the clusters for 2009, for reasons of comparability with the results obtained for 2006, the hierarchical cluster methodology was applied, along with the Chebyshev method for determining the distance between the elements, and the furthest-neighbour method for cluster generation.

As for the 2006 analysis, the first significant clusters were obtained by halting their generation at 12 clusters. ANOVA performed for the significance level of 95% (Table 3.7) allowed the conclusion that all four variables were significant in terms of cluster membership, because for all four, Sig. $F = 0.000 < \alpha = 0.05$ and consequently the null hypothesis (H_{0_1}) was rejected and the alternative hypothesis (H_{1_1}) was accepted.

The mean, minimum and maximum values and the 95% CIs for PA, PV, TLF and VABA for clusters A1–A4 for 2009 are shown in Table 3.8. The first four clusters included the 15 countries of clusters A1–A3 in 2006, but with a different distribution, plus the addition of Bulgaria, making 16 states in total.

Thus, instead of the seven countries that were grouped in cluster A1 in 2006, in 2009 they were grouped into two clusters: A1 and A4, respectively. Cluster A1 included three states (Belgium, Ireland and Austria) and cluster A4 included the other four (Czech Republic, Finland, Sweden and Norway). Cluster A2 retained the same composition (Estonia, Lithuania, Latvia, Luxembourg, Slovakia and Slovenia), while cluster A3 additionally includes Bulgaria, together with Hungary and Portugal.

Except for the VABA variable in cluster A1, the minimum and maximum limits of the 95% CIs for the mean values of the other variables presented in Table 3.8 had the same sign, which confirmed acceptance of the alternative hypothesis (H_{1_3}), with the mean values of the respective variables being statistically significant.

Table 3.7. ANOVA table for testing the statistical significance of PA, PV, TLF and VABA corresponding to 12 cluster groupings in 2009.

Variable		Sum of squares	df	Mean square	F	Sig.
PA	Between groups	936,905,942.2	11	85,173,267.5	624.1	0.00
	Within groups	1,637,748.3	12	136,479.0		
	Total	938,543,690.5	23			
PV	Between groups	2,148,056,582.8	11	195,277,871.2	686.4	0.00
	Within groups	3,414,080.3	12	284,506.7		
	Total	2,151,470,663.1	23			
TLF	Between groups	8,725,916.1	11	793,265.1	370.0	0.00
	Within groups	25,727.6	12	2,144.0		
	Total	8,751,643.6	23			
VABA	Between groups	1,207,005,030.3	11	109,727,730.0	740.8	0.00
	Within groups	1,777,420.0	12	148,118.3		
	Total	1,208,782,450.3	23			

Table 3.8. Mean, minimum and maximum values and 95% CIs of PA, PV, TLF and VABA for clusters A1–A4 in 2009. Note that clusters A1, A2 and A3 contain a total of 16 countries; at this point in cluster generation, the other eight countries are not included in the clusters.

Variable	Cluster	N	Mean	SD	SE	95% CI for mean LB	95% CI for mean UB	Min.	Max.
PA	A1	3	3248.50	447.88	258.58	2135.91	4361.09	2750.50	3618.30
	A2	6	444.62	235.88	96.30	197.08	692.15	153.60	750.90
	A3	3	1843.97	634.93	366.58	266.71	3421.22	1125.30	2328.90
	A4	4	1845.93	225.16	112.58	1487.64	2204.21	1527.50	2040.00
PV	A1	3	2298.92	861.61	497.45	158.55	4439.28	1378.00	3085.43
	A2	6	534.94	339.35	138.54	178.82	891.06	147.30	1004.70
	A3	3	2857.13	731.67	422.43	1039.55	4674.71	2016.76	3352.67
	A4	4	1590.29	307.06	153.53	1101.70	2078.89	1301.50	1933.61
TLF	A1	3	113.47	44.40	25.63	3.18	223.77	63.00	146.50
	A2	6	73.18	50.66	20.68	20.01	126.35	3.61	147.10
	A3	3	405.32	58.50	33.78	259.99	550.64	337.87	442.28
	A4	4	80.47	26.49	13.25	38.31	122.62	54.00	114.60
VABA	A1	3	1611.44	691.77	399.39	−107.00	3329.89	832.72	2154.87
	A2	6	243.13	137.51	56.14	98.82	387.43	80.35	433.60
	A3	3	1709.14	560.12	323.39	317.73	3100.56	1198.89	2308.46
	A4	4	911.55	181.03	90.52	623.48	1199.61	643.10	1026.17

A comparative analysis of the results presented in Table 3.2 and Table 3.8 demonstrated that the economic crisis triggered in 2008 had an important impact not only on the values of the analysed variables but also on the structuring of the 16 countries in clusters A1–A4 in 2009. A graphical image explaining the main features of these clusters is given in Fig. 3.7.

Cluster A1 for 2009, compared with 2006, was characterized by mean values of PA (€3248.5 million) higher than those registered in 2006, and also by

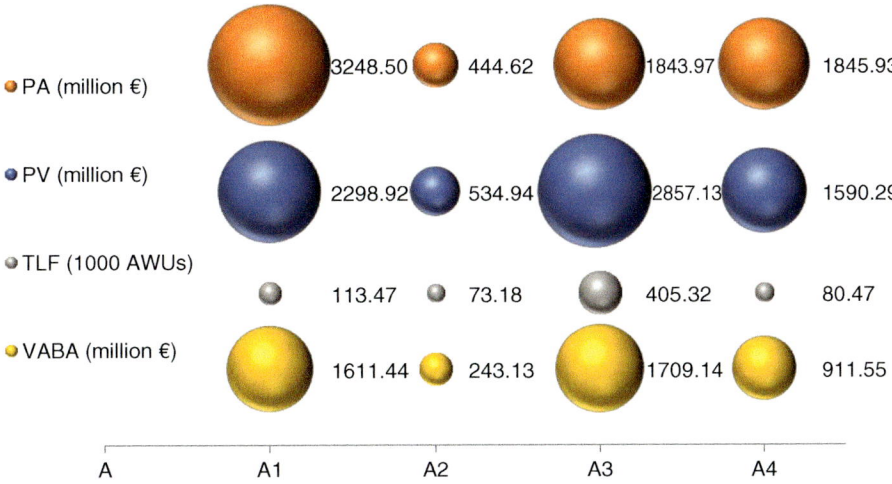

Fig. 3.7. Comparative presentation of the values of the variables PA, PV, TLF and VABA in clusters A1–A4 in 2009. (Based on information from Eurostat, 2017a,b.)

much lower values for TLF (29.1% lower); in 2009, the lowest value of PA was €2750.5 million in Austria and the highest was €3618.3 million in Belgium.

In addition, for cluster A1 in 2009, the mean value of PV was lower than the mean value of PA by about 30%. The 95% CI of the average value of PV was €158.55 million to €4439.28 million.

The range of values for PV in the three countries included in cluster A1 varied between a minimum of €1378.00 million in Ireland and a maximum of €3085.43 million in Belgium.

In 2009, compared with 2006, the significant decrease in the mean value of TLF in cluster A1 was mainly due to exclusion of the Czech Republic and Finland from this cluster, where values of TLF were above the mean value of 2006, and to a lesser extent to the reduction of activities as a result of the economic crisis. The lowest value of this indicator (in 1000 AWUs) was recorded in Belgium (63.0) and the highest in Austria (130.92).

The value of VABA in cluster A1 in 2009 was €1611.44 million compared with only €1362.21 million in 2006. The range of values of VABA varied among the countries that formed cluster A1 in 2009 from €832.72 million in Ireland to €2154.87 million in Austria. In terms of LPA in 2009, cluster A1 recorded values ranging from €5.6 million/1000 AWUs in Ireland to €29.31 million/1000 AWUs in Belgium.

For 2009, when compared with the other three clusters, cluster A2 was characterized by the lowest mean values of all four variables under investigation. It also maintained the same characteristic of this cluster in 2006, namely the very large spread of values of the variables around their mean, which led to coefficients of variation well above 30%. This provided reduced representability of the mean values of the four analysed variables.

The mean value of PA in 2009 in cluster A2 was €444.62 million. The 95% was €197.08 million to €692.15 million, and the range of values of PA in all six countries was from €153.60 million in Luxembourg to €750.90 million in Slovakia.

Unlike the situation in 2006, when the PV value in cluster A2 was sensibly equal to that of PA, in 2009 it was found that the mean value of PV (€534.94 million) was higher than that of PA by 20.03%. For the mean value of PV, the 95% CI was €178.82 million to €891.06 million. The range of the values for PV in the countries included in cluster A2 in 2009 was from €147.30 million in Luxembourg to €1004.70 million in Lithuania.

The mean TLF value (in 1000 AWUs) recorded in cluster A2 for 2009 was 73.18, with a 95% CI of €20.01 million to €126.35 million, and TLF values ranging from 3.61 in Luxembourg to 147.10 in Lithuania.

The mean value of VABA in cluster A2 in 2009 was €243.13 million compared with €284.09 million in 2006. The lowest value of this indicator was again recorded in Luxembourg (€80.35 million).

In terms of LPA, in 2009, cluster A2 recorded values ranging from €1.8 million/1000 AWUs in Latvia to €22.26 million/1000 AWUs in Luxembourg.

Cluster A3 consisted of three countries in 2009: Bulgaria, Hungary and Portugal. Unlike in 2006, when both PA and PV values in this cluster were relatively close, in 2009 the average PA value was €1843.97 million and PV was €2857.13 million (54.94% higher). In 2009, the lowest values for PA and PV were recorded in Bulgaria (€1125.30 million for PA and €2016.76 million for PV) and the highest in Portugal (€2328.90 million for PA and €3352.67 million for PV).

In 2009, the mean TLF value (in 1000 AWUs) in cluster A3 was 405.32, with the CI ranging between 259.99 and 550.64; the range of TLF values in the three states was from 337.87 in Portugal to 442.28 in Hungary.

The mean VABA value in cluster A3 for 2009 was €1709.14 million. For this indicator, the 95% CI was €317.73 million to €3100.56 million. The values of VABA in 2009 in the three states under consideration were €1846.74 million in Bulgaria, €1620.08 million in Hungary and €2308.46 million in Portugal. The LPA in cluster A3 in 2009 varied from €6.83 million/1000 AWUs in Portugal to €2.75 million/1000 AWUs in Bulgaria.

Cluster A4 was characterized by PA values approximately equal to those in cluster A3, with extreme values being recorded in the Czech Republic (€1527.5 million) and Norway (€2040 million). However, in terms of mean PV values, as well as of VABA, cluster A4 ranked third after clusters A3 and A1. In the case of PV, the lowest values were registered in Finland (€1301.5 million) and the Czech Republic (€1933.61 million), while for VABA, this was in the Czech Republic (€6431 million).

In cluster A4, LPA was between €5.61 million/1000 AWUs in the Czech Republic and €18.77 million/1000 AWUs in Norway.

For further comparison of the results derived from grouping the 24 states into four clusters based on the values of the PA, PV, TLF and VABA variables recorded in 2009 and 2006, the clustering process continued using the hierarchical cluster method to determine the clusters C1–C4. ANOVA for the cluster grouping performed at a significance level of 95% also led to the conclusion that all four variables were significant in terms of cluster membership (Table 3.9).

Table 3.9. ANOVA table for testing the statistical significance of PA, PV, TLF and VABA corresponding to the four cluster groupings in 2009.

Variable		Sum of squares	df	Mean square	F	Sig.
PA	Between groups	856,499,834.0	3.0	285,499,944.7	69.6	0.000
	Within groups	82,043,856.5	20.0	4,102,192.8		
	Total	938,543,690.5	23.0			
PV	Between groups	2,009,243,698.0	3.0	669,747,899.3	94.2	0.000
	Within groups	142,226,965.1	20.0	7,111,348.3		
	Total	2,151,470,663.1	23.0			
TLF	Between groups	8,265,760.1	3.0	2,755,253.4	113.4	0.000
	Within groups	485,883.5	20.0	24,294.2		
	Total	8,751,643.6	23.0			
VABA	Between groups	1,121,758,525.3	3.0	373,919,508.4	85.9	0.000
	Within groups	87,023,925.0	20.0	4,351,196.2		
	Total	1,208,782,450.3	23.0			

As in the previous analyses for all other clusters, the mean, minimum and maximum values of PA, PV, TLF and VABA as well as the 95% CIs for clusters C1–C4 in 2009 resulting from the clustering process were determined and are presented in Table 3.10.

The grouping obtained in 2009 (Table 3.10), although different from that for 2006 (Table 3.4), was also the result of the grouping within cluster C1 of the countries that previously formed clusters A1–A4. The other eight states initially not included in these clusters now formed three new clusters. Cluster C2 consists of four states (Germany, Spain, France and Italy), cluster C3 consisted of two states (the Netherlands and the UK) and cluster C4 comprised Romania and Poland. These aspects are also highlighted by the graphical representation of the values of the parameters analysed for clusters C1–C4 (Fig. 3.8).

The differences between the characteristics of the four clusters were significant. Cluster C1 was characterized by very low values of all the indicators analysed. This was due, on the one hand, to the fact that it encompassed countries with smaller areas than clusters C2–C4. On the other hand, cluster C1, in 2009, had a very high spread of values for all four variables in relation to their mean values, which led to very poor relevance of the mean values.

For cluster C2, the values of the coefficient of variation were less than 30% (23.28% for PA, 23.92% for PV, 29.79% for TLF and 26.28% for VABA) and consequently the mean values of the four variables had a good representability.

The mean value of PA in 2009 was €17,398.6 million. The lowest value of €13,419.5 million was recorded in Spain and the highest value of €21,476.7 million in France.

Regarding PV, the four countries in cluster C4 led to a mean value of €26,467.5 million, which was €6849.1 million higher than the sum of the mean values of PV in the other three clusters. The minimum value of PV in cluster C2 was recorded in Germany (€21,590 million) and the maximum in France (€35,463.2 million).

Another characteristic of cluster C2 was that the mean value of TLF was only 875.7 (in 1000 AWUs), whereas the average VABA was €19,515.8 million,

Table 3.10. Mean, minimum and maximum values and 95% CIs of PA, PV, TLF and VABA for clusters C1–C4 in 2009.

Variable	Cluster	N	Mean	SD	SE	95% CI for mean		Min.	Max.
						LB	UB		
PA	C1	16	1,583.1	1,102.0	275.5	995.8	2,170.3	153.6	3618.3
	C2	4	17,398.6	4,051.0	2,025.5	10,952.6	23,844.5	13,419.5	21,476.7
	C3	2	10,239.5	2,161.4	1,528.4	−9,180.1	29,659.0	8,711.1	11,767.8
	C4	2	6,060.5	3,150.4	2,227.7	−22,245.1	34,366.1	3,832.8	8,288.2
	Total	24	5,313.5	6,388.0	1,303.9	2,616.0	8,010.9	153.6	21,476.7
PV	C1	16	1,564.9	1,045.9	261.5	1,007.6	2,122.2	147.3	3,352.7
	C2	4	26,487.5	6,334.6	3,167.3	1,6407.8	36,567.2	21,590.0	35,463.2
	C3	2	9,536.0	2,327.1	1,645.5	−11,372.3	30,444.4	7,890.5	11,181.6
	C4	2	8,537.5	154.3	109.1	7,151.2	9,923.7	8,428.4	8,646.6
	Total	24	6,964.0	9,671.7	1,974.2	2,880.0	11,048.0	147.3	35,463.2
TLF	C1	16	144.8	136.5	34.1	72.1	217.6	3.6	442.3
	C2	4	857.7	255.5	127.8	451.1	1,264.3	532.2	1,149.0
	C3	2	218.1	92.2	65.2	−610.5	1,046.8	152.9	283.3
	C4	2	2,182.9	43.7	30.9	1,790.3	2,575.5	2,152.0	2,213.8
	Total	24	439.6	616.9	125.9	179.1	700.1	3.6	2,213.8
VABA	C1	16	941.7	722.1	180.5	556.9	1,326.5	80.4	2,308.5
	C2	4	19,515.8	5,128.0	2,564.0	11,356.0	27,675.5	13,245.0	25,795.1
	C3	2	7,684.5	402.7	284.7	4,066.8	11,302.3	7,399.8	7,969.3
	C4	2	5,705.9	388.7	274.9	2,213.1	9,198.6	5,431.0	5,980.7
	Total	24	4,996.3	7,249.5	1,479.8	1,935.1	8,057.5	80.4	25,795.1

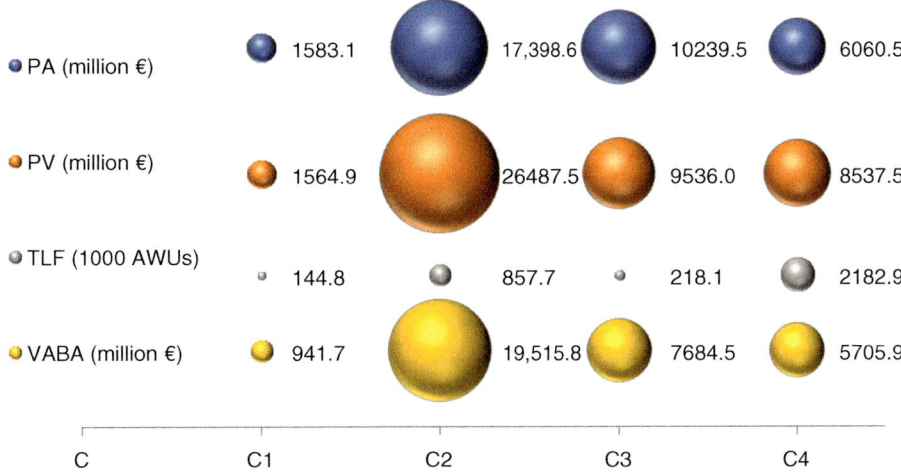

Fig. 3.8. Comparative presentation of the values of the variables PA, PV, TLF and VABA in clusters C1–C4 in 2009. (Based on information from Eurostat, 2017a,b.)

leading to high LPA in the countries of this cluster. Hence, in 2009, LPA was €24.89 million/1000 AWUs in Germany, €20.88 million/1000 AWUs in Spain, €23.89 million/1000 AWUs in France and €22.45 million/1000 AWUs in Italy.

Cluster C3 consisted of the Netherlands and the UK. The Netherlands had the lowest values for PA and TLF and the highest values for PV and VABA, producing a LPA value of €52.12 million/1000 AWUs (the highest of all the countries analysed).

Cluster C4 consisted of Romania and Poland, and was characterized, as in 2006, by a large TLF of 2152 (in 1000 AWUs) in Romania and 2213.8 in Poland. At the same time, VABA was €5431.0 million in Poland and €5980.7 million in Romania, which, for 2009, still led to some of the lowest LPA values of €2.45 million/1000 AWUs in Poland and €2.78 million/1000 AWUs in Romania.

3.5 Cluster Analysis Within the CAP in 2015

After 2009, PA and PV recorded increases both at the EU-28 level and in the case of the analysed countries, except for Bulgaria where PA showed a reduction of 16.98%. In 2015, PA was higher than in 2009, with values ranging from 1.82% in Romania to 58.31% in Ireland. In Bulgaria, however, PA dropped from €1125.3 million in 2009 to €934.2 million in 2015.

For PV, the largest percentage increase in production value was in Estonia, where it grew by 88.68% from €226.57 million in 2009 to €427.49 million in 2015, and the smallest in Finland, where the growth was only 3.87% from €1301.5 million in 2009 to €1351.9 million in 2015.

In the case of TLF, of the 24 states under analysis, five countries (Ireland, Lithuania, Hungary, Slovenia and the UK) had increases ranging from 1.5% in Slovenia (from 80.19 in 2009 to 81.4 in 2015, in 1000 AWUs) to 11.69% in

Ireland (from 146.5 in 2009 to 163.6 in 2015). In the other 19 countries analysed, the reductions ranged from 38.38% in Romania (from 2152 in 2009 to
1326 in 2015) to 0.55% in Luxembourg (from 3.61 in 2009 to 3.6 in 2015).

These developments led to a 23.36% increase in VABA in the EU-28, and of
the 24 states analysed, the highest increase of 187.73% was recorded in Ireland
(from €832.72 million in 2009 to €2396 million in 2015) and the lowest of
1.94% in Romania (from €5980.74 million in 2009 to €6097 million in 2015).

In order to determine the clusters for 2015, the hierarchical cluster methodology, Chebyshev and furthest-neighbour methods were also applied.

For 2015, the results of ANOVA in the case of a 12-cluster grouping and for a
significance level of 95% are presented in Table 3.11. Given that for the chosen
level of significance ($\alpha = 0.05$) the values of Sig. *F* were much smaller than α,
the null hypothesis H_{0_1} was rejected and the alternative hypothesis H_{1_1} was
accepted and consequently all four variables were significant from the point of
view of cluster membership.

For the first three significant clusters, A1–A3, the mean, minimum and
maximum values and 95% CIs of the PA, PV, TLF and VABA variables are shown
in Table 3.12. By analysing the limits of the 95% CIs for the mean values of the
variables under analysis (Table 3.12), the main result was that for cluster A1,
the mean of the TLF variable was not statistically significant (hypothesis H_{0_3} is
accepted).

The same conclusion applied to the PA, PV and VABA variables in cluster
A2. If we choose a 90% CI ($\alpha = 0.1$), the mean value of the TLF variable in cluster A1 becomes statistically significant (hypothesis H_{1_3} is accepted).

By comparing the cluster structures generated for 2015 with those generated for 2009, the resulting structure of cluster A1 was identical, that of cluster
A2 in 2015 comprised the countries in A4 without Hungary, and cluster A3 in
2015 included countries that in 2009 formed clusters A2 and A3. Altogether,
15 countries were included in the three clusters, with the other nine states still
forming single-element clusters.

Table 3.11. ANOVA table for testing the statistical significance of PA, PV, TLF and VABA
corresponding to 12 cluster groupings in 2015.

Variable		Sum of squares	df	Mean square	*F*	Sig.
PA	Between groups	1,278,275,870.1	11	116,206,897.3	114.6	0.000
	Within groups	12,167,995.4	12	1,013,999.6		
	Total	1,290,443,865.4	23			
PV	Between groups	2,943,709,353.5	11	267,609,941.2	322.4	0.000
	Within groups	9,962,182.9	12	830,181.9		
	Total	2,953,671,536.5	23			
TLF	Between groups	5,486,474.3	11	498,770.4	277.8	0.000
	Within groups	21,543.4	12	1,795.3		
	Total	5,508,017.7	23			
VABA	Between groups	1,895,474,695.1	11	172,315,881.4	602.7	0.000
	Within groups	3,431,106.4	12	285,925.5		
	Total	1,898,905,801.4	23			

In 2015, cluster A1 comprised three states: Belgium, Ireland and Austria. Compared with 2009, this cluster was characterized by higher values of PA (€4235.1 million compared with €3248.5 million in 2009, an increase of 30.37%), as well as of PV (€2831.8 million compared with €2298.92 million in 2009, an increase of 23.18%) and VABA (€2416.4 million compared with €1611.44 million in 2009, an increase of 49.96%). These trends were accompanied by a very small increase in TLF from 113.47 (in 1000 AWUs) in 2009 to 113.6 in 2015 .

A comparative presentation of the mean values recorded by the variables analysed in the three clusters is illustrated in Fig. 3.9.

With respect to PA, cluster A1 in 2015 showed the lowest and highest values in Austria (€3254.8 million) and Ireland (€5345.8 million), respectively.

Table 3.12. Mean, minimum and maximum values and CIs of PA, PV, TLF and VABA for clusters A1–A3 in 2015. Note that clusters A1–A3 each contain a total of 15 states; at this point in cluster generation, the other nine states are not included.

| Variable | Cluster | N | Mean | SD | SE | 95% CI for mean | | Min. | Max. |
						LB	UB		
PA	A1	3	4235.1	1051.6	607.1	1622.8	6847.4	3254.8	5345.8
	A2	2	1810.6	1239.4	876.4	−9325.1	12946.3	934.2	2687.0
	A3	10	1223.0	967.3	305.9	531.0	1914.9	200.0	2670.5
PV	A1	3	2831.8	1017.1	587.2	305.2	5358.5	1772.3	3800.4
	A2	2	3103.2	802.7	567.6	−4108.7	10315.0	2535.6	3670.7
	A3	10	1331.5	897.5	283.8	689.5	1973.5	159.8	2813.5
TLF	A1	3	113.6	53.8	31.0	−19.9	247.2	56.8	163.6
	A2	2	269.8	9.4	6.6	185.7	353.8	263.2	276.4
	A3	10	67.7	41.7	13.2	37.9	97.6	3.6	150.8
VABA	A1	3	2416.4	293.0	169.2	1688.6	3144.2	2134.2	2719.1
	A2	2	1911.2	728.4	515.1	−4633.2	8455.6	1396.2	2426.3
	A3	10	734.1	550.6	174.1	340.2	1128.0	90.0	1683.5

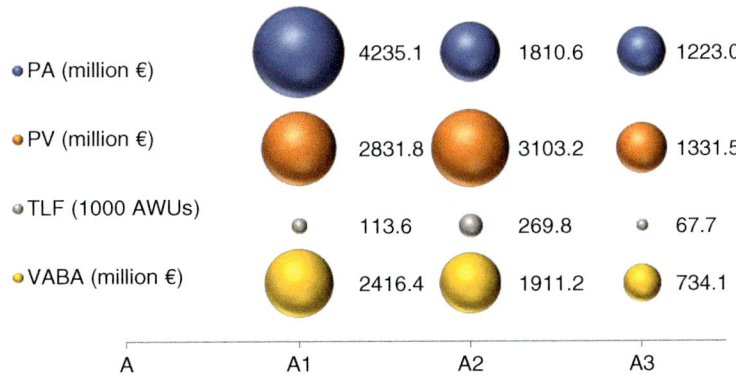

Fig. 3.9. Comparative presentation of the values of the variables PA, PV, TLF and VABA in clusters A1–A3 in 2015. (Based on information from Eurostat, 2017a,b.)

For PV, the highest value of €3800.4 million was recorded in Belgium and the lowest of €1772.3 million in Ireland.

For cluster A1 in 2015, the average value of the VABA indicator showed very good representability (the coefficient of variation was 12.12%), with the spread of values being very low. The lowest and highest values of VABA were recorded in Belgium (€2134.2 million) and Austria (€2719.1 million), respectively.

As far as TLF is concerned, the spread of values was much higher than the average value of the cluster. The lowest value of 56.8 (in 1000 AWUs) was recorded in Belgium and the highest of 163.6 was in Ireland.

These values led to fairly large differences in LPA, which in 2015 varied from €14.64 million/1000 AWUs in Ireland to €22.56 million/1000 AWUs in Austria and €37.59 million/1000 AWUs in Belgium.

In 2015, cluster A2 comprised only two countries (Bulgaria and Portugal) compared with three in 2006 (Bulgaria, Portugal and Hungary). Due to the significant changes found in the values of the analysed variables in Hungary, the distance between Hungary and the other two states increased, which led to the exclusion of Hungary from this cluster. Under these circumstances, the mean value of PA dropped from €1845.93 million in 2009 in cluster A4 to €1810.6 million recorded in 2015 in cluster A2.

In contrast, several increases were recorded for PV (from €1590.29 million in 2009 to €3103.2 million in 2015) and VABA (from €911.55 million in 2009 to €1911.2 million in 2015). In Bulgaria in 2015, the value of PV was €2535.6 million compared with €2016.7 million in 2009, while in Portugal it reached €3670.7 million compared with €3352.67 million in 2009.

At the same time, both countries faced reductions in TLF, from 337.87 (in 1000 AWUs) in 2009 to 263.2 in 2015 in Portugal, and from 435.8 in 2009 to 276.4 in 2015 in Bulgaria.

Developments in VABA and the size of the TLF led to an increase in LPA in Bulgaria from €2.75 million/1000 AWUs in 2009 to €5.05 million/1000 AWUs in 2015, while in Portugal, LPA increased from €6.83 million/1000 AWUs in 2009 to €9.22 million/1000 AWUs in 2015.

Cluster A3 included ten countries that in 2009 belonged to clusters A2 and A3; in this cluster, compared with clusters A1 and A2, the smallest values were found of the variables analysed.

In 2015, in cluster A3, the mean value of PA was €1223.0 million, within a 95% CI of €531.0 million to €1914.9 million. In this cluster, the highest value of PA was recorded in Sweden (€2670.5 million) and the lowest in Luxembourg (€200.0 million).

For PV, the mean value (€1331.5 million) was 2.33 times lower than that in cluster A2 and 2.13 times lower than that in cluster A1; the lowest value of PV was recorded in Luxembourg (€159.8 million) and the highest in Sweden (€2813.5 million).

Concerning VABA, the mean value in cluster A3 was €734.1 million, and the lowest and highest values were found in the same countries: Luxembourg (€90 million) and Sweden (€1683.5). In Luxembourg, due to its reduced area, the minimum TLF (3.6, in 1000 AWUs) was also recorded, while the maximum value is no longer recorded in Sweden but in Lithuania was 150.8.

In terms of LPA, the differences between countries within this cluster were significant. While Latvia, Lithuania, Slovenia, Slovakia and Finland recorded values below €10 million/1000 AWUs (the lowest value was €3.26 million/1000 AWUs in Latvia), Luxembourg, Sweden and Norway recorded values above €25 million/1000 AWUs (the highest value was €31.48 million/1000 AWUs in Norway).

By continuing to use the hierarchical cluster method for 2015, we also obtain four clusters, C1–C4, that included all states in the analysis. The results of the ANOVA for this group for a significance level of 95% (Table 3.13) also led to the conclusion that all four variables were significant from the point of view of cluster membership (the H_{0_1} hypothesis was rejected and hypothesis H_{1_1} was accepted).

The mean, minimum and maximum values, as well as the 95% CIs of the PA, PV, TLF and VABA variables for clusters C1–C4 corresponding to 2015 are presented in Table 3.14. A comparative presentation of the mean values of the variables analysed is illustrated in Fig. 3.10.

Cluster C1 contained, as in the case of 2009, 16 countries and was formed by including clusters A1, A2 and A3, plus Hungary. This provided the possibility of a good comparison of cluster C1 evolution during 2009–2015.

In 2015, the average values of PA, PV and VABA were higher than in 2009 by 22.48% (30.12% and 24.54%, respectively). At the same time, the average TLF decreased from 218.1 (in 1000 AWUs) in 2009 to 127 in 2015 (corresponding to a decrease of 11.8%).

In 2015, the mean value of PA in C1 was €1939.1 million, with the lowest and highest values found in Luxembourg (€200.0 million) and Ireland (€5345.8 million), respectively. The cluster's mean PV value was €2035.6 million, with the lowest value (€159.8 million) also recorded in Luxembourg and the highest (€4552.9 million) in Hungary.

For VABA, the lowest level was also recorded in Luxembourg (€90 million) and the highest (€2797.6 million) again in Hungary.

Table 3.13. ANOVA table for testing the statistical significance of PA, PV, TLF and VABA corresponding to four cluster groupings in 2015.

Variable		Sum of squares	df	Mean square	*F*	Sig.
PA	Between groups	1,108,809,944.7	3	369,603,314.9	40.7	0.000
	Within groups	181,633,920.8	20	9,081,696.0		
	Total	1,290,443,865.4	23			
PV	Between groups	2,654,702,737.2	3	884,900,912.4	59.2	0.000
	Within groups	298,968,799.2	20	14,948,440.0		
	Total	2,953,671,536.5	23			
TLF	Between groups	4,798,206.3	3	1,599,402.1	45.1	0.000
	Within groups	709,811.4	20	35,490.6		
	Total	5,508,017.7	23			
VABA	Between groups	1,790,126,334.4	3	596,708,778.1	109.7	0.000
	Within groups	108,779,467.1	20	5,438,973.4		
	Total	1,898,905,801.4	23			

Table 3.14. Mean, minimum and maximum values and 95% CIs of PA, PV, TLF and VABA for clusters C1–C4 in 2015.

Variable	Cluster	N	Mean	SD	SE	95% CI for mean		Min.	Max.
						LB	UB		
PA	C1	16	1,939.1	1,492.8	373.2	1,143.7	2,734.5	200.0	5,345.8
	C2	4	16,296.2	5,319.2	2,659.6	7,832.2	24,760.1	10,476.1	23,381.0
	C3	2	20,504.3	6,536.4	4,622.0	−38,223.2	79,231.7	15,882.3	25,126.2
	C4	2	7,112.1	4,539.1	3,209.7	−33,670.4	47,894.5	3,902.4	10,321.7
	Total	24	6,310.1	7,490.4	1,529.0	3,147.2	9,473.0	200.0	25,126.2
PV	C1	16	2,035.6	1,302.3	325.6	1,341.7	2,729.5	159.8	4,552.9
	C2	4	18,592.3	8,076.2	4,038.1	5,741.3	31,443.3	10,461.5	25,932.1
	C3	2	36,558.3	8,723.5	6,168.5	−41,819.7	114,936.3	30,389.8	42,726.8
	C4	2	10,041.6	1,325.6	937.3	−1,868.5	21,951.7	9,104.3	10,979.0
	Total	24	8,339.1	11,332.3	2,313.2	3,553.9	13,124.3	159.8	42,726.8
TLF	C1	16	127.0	120.7	30.2	62.7	191.3	3.6	474.3
	C2	4	435.4	284.5	142.3	−17.3	888.1	144.6	802.8
	C3	2	941.5	248.3	175.6	−1,289.4	3,172.4	766.0	1,117.1
	C4	2	1,631.6	432.1	305.6	−2,250.8	5,513.9	1,326.0	1,937.1
	Total	24	371.7	489.4	99.9	165.0	578.3	3.6	1,937.1
VABA	C1	16	1,325.7	943.5	235.9	822.9	1,828.4	90.0	2,797.6
	C2	4	13,271.9	5,396.5	2,698.3	4,684.8	21,858.9	9,637.5	21,223.3
	C3	2	30,790.6	2,696.4	1,906.7	6,564.0	55,017.2	28,883.9	32,697.3
	C4	2	6,725.3	888.5	628.3	−1,257.6	14,708.2	6,097.0	7,353.6
	Total	24	6,222.1	9,086.3	1,854.7	2,385.3	10,058.9	90.0	32,697.3

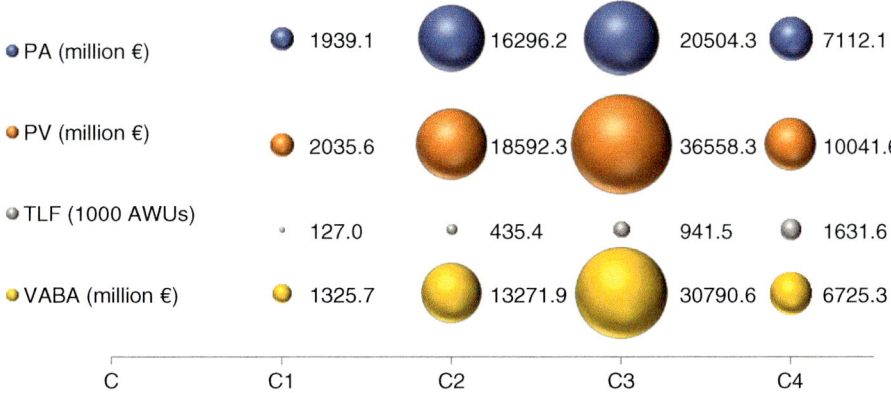

Fig. 3.10. Comparative presentation of the values of the variables PA, PV, TLF and VABA in clusters C1–C4 in 2015. (Based on information from Eurostat, 2017a,b.)

In parallel with the increase in average VABA, between 2009 and 2015, there was also an 11.8% reduction in TLF, which resulted in an increase in LPA for this cluster, from €6.51 million/1000 AWUs in 2009 to €10.44 million/1000 AWUs in 2015.

The highest LPA in cluster C1 was recorded in Norway (€31.48 million/1000 AWUs) and the lowest in Latvia (€3.26 million/1000 AWUs).

In 2015, cluster C2 included, as in 2009, four countries. However, unlike in 2009, in addition to Germany and Spain, which retained cluster membership, instead of France and Italy, which left the cluster, the Netherlands and the UK, which in 2009 formed a separate cluster (cluster C3), joined cluster C2.

In cluster C2, Spain had the highest values of PV (€25,932.1 million), TLF (802.8, in 1000 AWUs) and VABA (€21,223.3 million), while Germany showed the highest PA value (€23,381 million). The lowest values of PA (€10,476.1 million), TLF (144.6, in 1000 AWUs) and VABA (€9637.5 million) were recorded in Germany, while the lowest value of PV was in the UK (€10,461.5 million).

LPA in cluster C2 varied between €24.12 million/1000 AWUs in Germany and €66.66 million/1000 AWUs in the Netherlands (the highest productivity recorded in 2015 among all of the countries analysed).

For 2015, cluster C3 consists of France and Italy, which in 2009 formed part of cluster C2.

In 2015, the maximum values for PA (€25,126.2 million) and PV (€42,726.8 million) and the minimum values for TLF (766, in 1000 AWUs) and VABA (€28,883.9 million) were in France, resulting in a LPA value of €37.71 million/1000 AWUs. At the same time, the maximum values for TLF (1117.1, in 1000 AWUs) and VABA (€32,697.3 million) were found in Italy, where the lowest values of PA (€15,882.3 million) and PV (€30,389.8 million) were also found.

LPA in cluster C3 reached €29.27 million/1000 AWUs in Italy and €37.71 million/1000 AWUs in France.

Cluster C4 in 2015 was formed by Romania and Poland and was characterized, as in previous years, by high TLF values (1326 in Romania and 1937.1 in Poland, in 1000 AWUs). VABA was €7353.6 million in Poland and €6097 million in Romania.

Unlike in 2009, when LPA was €2.45 million/1000 AWUs in Poland and €2.78 million/1000 AWUs in Romania, in 2015 there were significant increases, especially in Romania, reaching €4.6 million/1000 AWUs, while in Poland this indicator reached €3.8 million/1000 AWUs.

Notes

[1] This chapter is based on research carried by Andrei (2016).
[2] Utilized agricultural area (UAA) includes arable land (including temporary pastures, fallow land, greenhouse crops and family gardens), the areas always under grass cover, and permanent crops (e.g. vineyards, orchards) (European Network for Rural Development, 2010).

4 Correlations, Trends and Realities in the European Agri-food Model[1]

4.1 Aspects of the European Agri-food Model and Agricultural Transformations

The European agricultural sector and European agriculture have generally undergone dynamic transformations throughout their existence, evolving with significant fluctuations, and resulting in the transformation and evolution of the EU as a whole. At the European level, agriculture does not imply and is not limited to food production and food safety and security, but refers to a wide array of issues, ranging from land and farm management, environmental and rural landscapes conservation and the rational use of water resources to the production of essential public goods.

As noted in the literature (Anderson and Tyers, 1993; Andrei and Begalli, 2015; Nazzaro and Marotta, 2016; Buller and Hoggart, 2017), the European agricultural model practically defines a social contract between farmers and society, whereby the former ensure sustainable food production at the highest standards, satisfying the conditions of food safety and security but also the demands on environmental protection, public goods production, and landscape and habitat conservation. The European agricultural model embraces all of these aspects and refines them, increasing the degree of convergence between the agricultural sectors of the EU Member States.

The challenges faced by European agriculture in general, and the fact that the agricultural sector is affected by climate change and the pressure of the population on the level of agricultural production, imply, above all, a process of adapting to new demands but also the need for flexibility of specific budget availability. Although the CAP has generated and continues to generate well-being for European citizens, contributing significantly to the development of the agricultural sector and European agriculture, the persistence of income inequality among farmers in different European countries and the deepening of regional disparities remain issues to be resolved. Often, agricultural incomes are insufficient to ensure a decent living for farmers, and their growth rate is in most cases below the national average.

In this context, the CAP is the only European policy capable of supporting the sustainable exploitation of agricultural potential and ensuring the proper development of family farms, cooperatives and agricultural societies, while setting them on the basis of economic and social efficiency.

Starting from the results and conclusions drawn in the previous chapter from the cluster analysis of the 24 states in the years 2006, 2009 and 2015 according to the four variables of animal production (PA), vegetal production (PV), total labour force input in the agri-food sector (TLF) and gross value added in the agri-food sector (VABA), in order to analyse the correlations between these variables and to characterize trends in the agri-food industry, the eight countries that in 2009 and 2015 were included in clusters C2–C4 were selected for analysis, as these clusters presented mean values of parameters that were significantly different from the 16 countries included in cluster C1.

For 2015, the eight countries were grouped into clusters as follows: cluster C2 comprised Germany, Spain, the Netherlands and the UK; cluster C3 consisted of France and Italy; and cluster C4 contained Poland and Romania.

4.2 Characteristics of the Evolution of the Agricultural Sector in Germany, Spain, the Netherlands and the UK

In cluster C2, the mean values of PA, PV and VABA were statistically significant for a 95% CI (the null hypothesis H_{0_2} was rejected and the alternative hypothesis H_{1_2} was accepted), and the mean value of TLF was statistically significant for a 90% CI.

As the range of the CI and the test statistics depend on the standard error (SE), which in turn is determined by the scattering of values around the mean, a fair conclusion that can be drawn is that, besides the similar characteristics that led to their grouping into this cluster, there were also elements or aspects that differentiated them.

For Germany, the main features of the mean values of the four indicators over the period 2006–2015 are presented in Table 4.1.

Table 4.1. Characteristics of the mean values of the evolution of VABA, PA, PV and TLF in Germany, 2006–2015.

Variable	Mean	SD	SE	*t_statistic*	95% CI for mean LB	95% CI for mean UB	V (%)
VABA	15,873.97	2,758.07	872.18	18.200	13,901.11	17,846.84	17.37
PA	23,043.88	2,459.78	777.85	29.625	21,284.38	24,803.38	10.67
PV	24,997.76	3,316.62	1,048.81	23.834	22,625.35	27,370.16	13.27
TLF	525.56	22.20	7.02	74.860	509.68	541.44	4.22

Taking into account that for all the mean values of the four variables, the values of the *t_statistic* were higher than the critical value $t_{\alpha=0.05,9}$ for the bilateral test ($t_{critic}=2262$), the null hypothesis H_{0_2} was rejected and the alternative hypothesis H_{1_2} was accepted, indicating that all mean values were statistically significant.

In addition, the values of the coefficient of variation (*V*) showed that, for the variables PA, PV and TLF, the mean values had a very good representability, while for VABA the representability of the mean was good.

The higher value of *V* for VABA in relation to the values of the other *V* values highlighted the fact that the scattering of the values recorded between 2006 and 2015 around its mean was greater than for the other variables.

The ways in which the values of the four variables evolved around the mean and the trend are shown in Fig. 4.1. For a better graphical comparison, instead of TLF (given in absolute numbers as 1000 AWUs), the variable TLF_h, obtained using the transformation TLF_h = TLF × 10, with the unit of measurement as 100 AWUs, was used.

The first observation seen from the graphical representations of the variable values is the fact that their slopes are different. The largest slope, which corresponds to an average annual increase of €800.90 million, was recorded in the evolution of PV. This situation showed a more pronounced increase in the value of PV compared with PA, for which the annual average increase was €641.13 million. Unlike the values of PA and PV, VABA grew with an average annual increase of only €180.60 million, while the TLF decreased on average by 7.53 (in 1000 AWUs).

A second observation relates to the impact that the economic crisis had on the values of the variables. This affected both PA and PV values, which in 2009

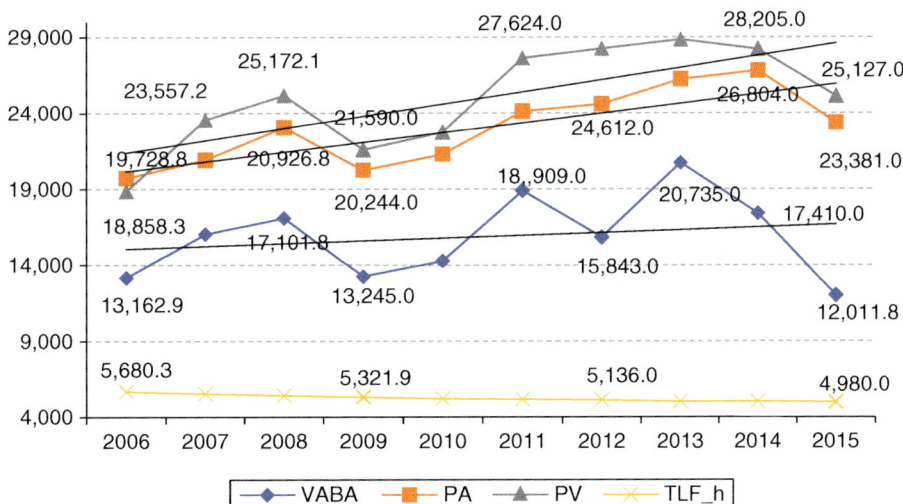

Fig. 4.1. Evolution of the variables VABA, PA, PV and TLF_h in Germany, 2006–2015. The *y*-axis units are million € for VABA, PA and PV, and 100 AWUs for TLF_h. (Based on information from Eurostat, 2017a,b.)

decreased by 12.47% and 14.23%, respectively, compared with 2008, as well VABA, which showed a significant reduction of 22.55%. With regard to TLF, whose trend was to decline throughout the period, the reduction was only 2.0%.

A third observation concerned the amplitude of the oscillations of the values of the variables around the trend. Thus, while after 2009 the value of PA and PV grew relatively steadily until 2014, VABA oscillated strongly.

Finally, a decline was seen in the values of both PV and PA, and especially VABA, which fell by 42.07% between 2013 and 2015.

In order to show the strength of the correlations between the four variables, Pearson parametric correlation coefficients were determined. These tests were performed using the *t_bilateral* test for a significance threshold of $\alpha = 0.05$; the results obtained are shown in Table 4.2.

Considering that for the correlation between VABA and TLF, the value Sig. (two-tailed) = $0.438 > \alpha = 0.05$, the null hypothesis H_{0_4} was accepted and consequently the value of the correlation coefficient was not statistically significant. In all other cases, the values of the correlation coefficients were statistically significant. The strongest direct correlation existed between PA and PV ($r = 0.943$). Direct but medium-strength correlations also occurred between the variable VABA and the PA and PV variables. In addition, there were two inverse correlations of quite strong intensity between the TLF variable and the PA and PV variables.

Starting from the observation concerning the correlations between the variable VABA and the variables PA ($r = 0.689$) and PV ($r = 0.731$), three regression models were tested:

$$VABA_{D1} = a_{10} + a_{11} \times PA_D + a_{12} \times PV_D + \varepsilon_1$$
$$VABA_{D2} = a_{20} + a_{21} \times PA_D + \varepsilon_2 \qquad (10)$$
$$VABA_{D3} = a_{30} + a_{31} \times PV_D + \varepsilon_3$$

The results obtained for the models in Eqns 10 are presented in Table 4.3. From the three models tested for the $VABA_{D1}$ model, as $F = 4.0015 < F_{0.05;1;9} = 5.12$, the null hypothesis H_{0_5} was accepted and consequently this model was

Table 4.2. Values of the correlation coefficients between VABA, PA, PV and TLF in Germany during 2006–2015, and the significance thresholds.

Variable		VABA	PA	PV	TLF
VABA	Pearson correlation	1	0.689*	0.731*	−0.277
	Sig. (two-tailed)		0.028	0.016	0.438
PA	Pearson correlation	0.689*	1	0.943**	−0.778**
	Sig. (2-tailed)	0.028		0.000	0.008
PV	Pearson correlation	0.731*	0.943**	1	−0.763*
	Sig. (two-tailed)	0.016	0.000		0.010
TLF	Pearson correlation	−0.277	−0.778**	−0.763*	1
	Sig. (two-tailed)	0.438	0.008	0.010	

*Correlation is significant at the 0.05 level (two-tailed).
**Correlation is significant at the 0.01 level (two-tailed).

Table 4.3. Results of the testing of the model described in Eqns 10, parameter values, test results and CIs.

| Model | F | Sig. F | R^2 | Coefficient | | t | Sig. | 95% CI for B | |
				Name	B			LB	UB
$VABA_{D1}$	4.0051	0.069	0.533	a_{10}	666.347	0.09	0.93	−16,474.54	17,807.24
				a_{11}	0.004	0.00	1.00	−2.05	2.05
				a_{12}	0.605	0.94	0.38	−0.92	2.13
$VABA_{D2}$	7.2300	0.027	0.474	a_{20}	−1,928.602	−0.29	0.78	−17,274.43	13,417.22
				a_{21}	0.773	2.69	0.03	0.11	1.44
$VABA_{D3}$	9.1545	0.016	0.533	a_{30}	688.162	0.14	0.90	−10,977.11	12,353.44
				a_{31}	0.607	3.03	0.02	0.14	1.07

not statistically significant. The same conclusion was reached also considering that Sig. $F = 0.069 > \alpha = 0.05$.

For the $VABA_{D2}$ and $VABA_{D3}$ models, the null hypothesis H_{0_5} was rejected and the alternative hypothesis H_{1_5} was accepted, both models being statistically significant.

The $VABA_{D2}$ model describes the functional link between VABA and the value of PA in Germany over the period 2006–2015. As sign (LB) = sign (UB), the significance level chosen for parameter a_{21} led to rejection of the null hypothesis H_{0_6} and acceptance of the alternative hypothesis H_{1_6}, and consequently this model was statistically significant. The regression model had the following form:

$$VABA_{D2} = 1928.602 + 0.773 \times PA_D + \varepsilon_2 \tag{11}$$

Model (11), together with the CI of the regressor (a_{21}), highlighted that the increase in the value of PA by €1 million would lead to an increase in VABA in Germany by a value of between €0.11 million and €1.44 million, the most likely increase being by €0.773 million.

The $VABA_{D3}$ model describes the functional connection between VABA and the value of PV. In addition, parameter a_{31} was statistically significant (sign (LB) = sign (UB)), so the regression model is:

$$VABA_{D3} = 688.162 + 0.607 \times PV_D + \varepsilon_3 \tag{12}$$

Model (12), together with the CI of the regressor (a_{31}), indicate that the increase in the value of PV by €1 million would lead to an increase in VABA in Germany by a value ranging between €0.14 million and €1.07 million, the most likely increase being €0.607 million.

Table 4.4 shows the mean values of VABA, PA, PV and TLF for the studied period for Spain. For all the mean values, $t\_statistic > t_{\alpha\_2;9} = 2.262$ led to rejection of the null hypothesis H_{0_2} and the alternative hypothesis H_{1_2} was accepted, with all the mean values being statistically significant. In addition, the values of V indicated that, for all four variables, the mean values had a very good representativity.

The ways in which the values of the four variables have evolved around the mean and the trend are shown in Fig. 4.2. As for Fig. 4.1, the TLF_h variable is used for a better graphical comparison. The results showed that the slopes of the variables analysed were different for Spain, as occurred for Germany.

The highest slope, corresponding to an annual average increase of €327.66 million, was for the evolution of the value of PA, followed by the slope showing the evolution of the value of PV, with an annual average increase of €189.35 million.

Unlike in Germany, where the increase in VABA was positive, in Spain between 2006 and 2015 there was a regression (−€12.0 million per year). At the same time, unlike Germany, TLF recorded an average reduction of 24.9 (in 1000 AWUs).

The impact of the economic crisis on the four variables in Spain, compared with that in Germany, was much lower. Although in 2009 compared with 2008 there were reductions of 12.6% for PV compared with 14.23% in Germany, the

Table 4.4. Characteristics of the mean values of the evolution of VABA, PA, PV and TLF in Spain during 2006–2015.

Variable	Mean	SD	SE	t_statistic	95% CI for mean		V (%)
					LB	UB	
VABA	21,165.02	1,189.27	376.08	56.28	20,314.32	22,015.71	5.62
PA	14,697.04	1,183.12	374.14	39.28	13,850.74	15,543.33	8.05
PV	24,566.23	1,519.02	480.36	51.14	23,479.66	25,652.80	6.18
TLF	917.14	78.48	24.82	36.96	861.01	973.28	8.56

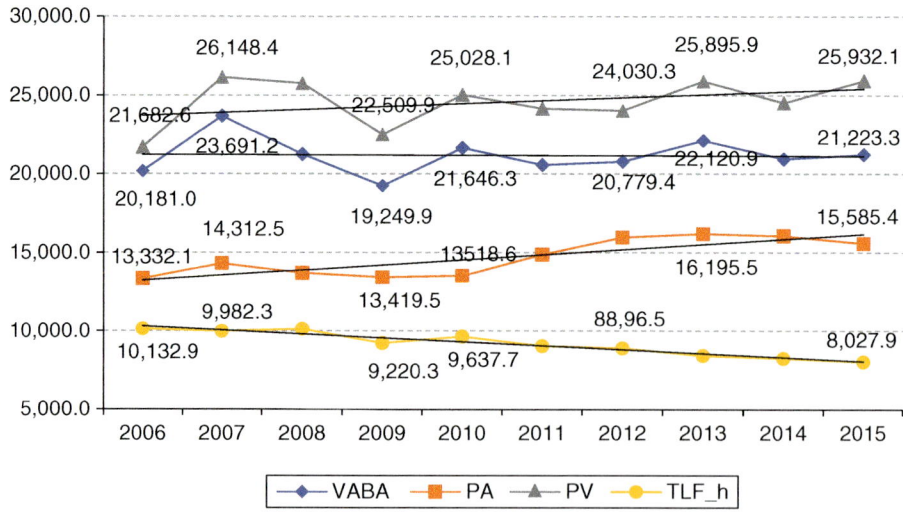

Fig. 4.2. Evolution of the variables VABA, PA, PV and TLF_h, in Spain, 2006–2015. See Fig. 4.1 for *y*-axis units. (Based on information from Eurostat, 2017a,b.)

impact on PV value was −1.9% (versus −12.25% in Germany) and on VABA, where the decrease was 9.4% compared with 22.56% in Germany.

As a significant difference between the evolutions of the four variables in Germany and in Spain, the significant correlations between the variables were much lower, as can be seen from Table 4.5.

During the analysed period, only two significant correlations were seen: a direct correlation between VABA and PV, and an inverse correlation between PA and TLF. It is important to note that the values of the correlation coefficients were statistically significant for a 99% CI.

Considering that there were no significant correlations between VABA and PA, or between VABA and TLF, and the values of the Pearson correlation coefficients between them were not statistically significant, and for Spain, a single model was tested:

$$VABA_E = a_0 + a_1 \times PV_E + \varepsilon \tag{13}$$

The results of testing model (13) are shown in Table 4.6. As $F = 13.626 > F_{0.05;1;9} = 5.12$, the null hypothesis H_{0_5} was rejected and the alternative hypothesis H_{1_5} was accepted, and consequently model (13) was statistically significant.

The $VABA_E$ model describes the functional link between VABA and the value of the Spanish PV in the period 2006–2015. For the significance level chosen, the parameter a_1 was statistically significant (sign(LB) = sign (UB)), and thus the regression was in the form:

$$VABA_E = 5898.062 + 0.168 \times PV_E + \varepsilon \tag{14}$$

Table 4.5. Values of the correlation coefficients between VABA, PA, PV and TLF in Spain, 2006–2015, and the significance thresholds.

Variable		VABA	PA	PV	TLF
VABA	Pearson correlation	1	0.217	0.794**	0.109
	Sig. (two-tailed)		0.547	0.006	0.764
PA	Pearson correlation	0.217	1	0.408	−0.836**
	Sig. (two-tailed)	0.547		0.242	0.003
PV	Pearson correlation	0.794**	0.408	1	−0.239
	Sig. (two-tailed)	0.006	0.242		0.507
TLF	Pearson correlation	0.109	−0.836**	−0.239	1
	Sig. (two-tailed)	0.764	0.003	0.507	

**Correlation is significant at the 0.01 level (two-tailed).

Table 4.6. Results of model (13) testing, parameter values, test results and their CIs.

Model	F	Sig.F	R^2	Coefficients				95% CI for B	
				Name	B	t	Sig.	LB	UB
$VABA_E$	13.626	0.006	0.630	a_0	5,898.062	1.42	0.192	−3,655.652	15,451.775
				a_1	0.168	3.69	0.006	0.233	1.010

Model (14), together with the CI of the regressor (a_1) indicates that with an increase of €1 million in the value of PV, the value of VABA in Spain would increase by a value of between €0.233 million and €1010 million, the most likely value being €0.168 million.

For the Netherlands, the characteristics of the mean values of VABA, PA, PV and TLF over the period 2006–2015 are presented in Table 4.7. For all the mean values, $t\_statistic > t_{\frac{\alpha}{2};9} = 2.262$, so the null hypothesis H_{0_2} was again rejected and the alternative hypothesis H_{1_2} was accepted, all the mean values being statistically significant.

In addition, considering the values of V, the resulting effect was that for, all four variables, the mean values had a very good representability. The evolution of the variables VABA, PA, PV and TLF_h and their trends are shown in Fig. 4.3.

The highest slope, corresponding to an annual average increase of €296.95 million, was recorded for PA. The next highest, with an average annual increase of €172.94 million, was for PV.

Table 4.7. Characteristics of the mean values of the evolution of VABA, PA, PV and TLF in the Netherlands, 2006–2015.

Variable	Mean	SD	SE	t_statistic	95% CI for mean LB	95% CI for mean UB	V (%)
VABA	21,165.02	1,189.27	376.08	56.28	20,314.32	22,015.71	5.62
PA	14,697.04	1,183.12	374.14	39.28	13,850.74	15,543.33	8.05
PV	24,566.23	1,519.02	480.36	51.14	23,479.66	25,652.80	6.18
TLF	917.14	78.48	24.82	36.96	861.01	973.28	8.56

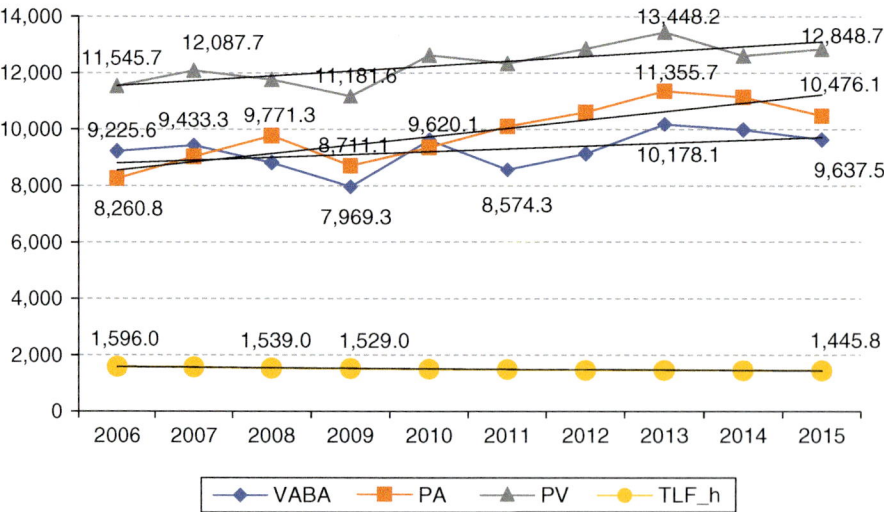

Fig. 4.3. Evolution of variables VABA, PA, PV and TLF_h, in the Netherlands, 2006–2015. See Fig. 4.1 for *y*-axis units. (Based on information from Eurostat, 2017a,b.)

Table 4.8. Values of the correlation coefficients between VABA, PA, PV and TLF in the Netherlands, 2006–2015, and their significance thresholds.

Variable		VABA	PA	PV	TLF
VABA	Pearson correlation	1	0.566	0.767**	−0.379
	Sig. (two-tailed)		0.088	0.010	0.280
PA	Pearson correlation	0.566	1	0.843**	−0.882**
	Sig. (two-tailed)	0.088		0.002	0.003
PV	Pearson correlation	0.767**	0.843**	1	−0.761*
	Sig. (two-tailed)	0.010	0.002		0.011
TLF	Pearson correlation	−0.379	−0.882**	−0.761*	1
	Sig. (two-tailed)	0.280	0.003	0.011	

*Correlation is significant at the 0.05 level (two-tailed).
**Correlation is significant at the 0.01 level (two-tailed).

In the analysed period, VABA in the Netherlands increased annually, on average, by €101.92 million. During the same period, there was also a slight decrease in TLF from 159.6 to 144.6 (in 1000 AWUs), with an annual average of 1.69.

The strength of the links between the four variables is highlighted by the Pearson correlation coefficient values (Table 4.8). For the significance threshold of $\alpha = 0.05$ between VABA and each of the other three variables, only the correlation with PV was statistically significant.

In addition, a significant direct correlation was observed between PA and PV (0.767) and an inverse correlation between PA and TLF (−0.882) for a significance threshold of $\alpha = 0.01$ and between PV and TLF (−0.761) for a significance threshold $\alpha = 0.05$.

Taking into account the values of the correlation coefficients in the case of the Netherlands, a single model was tested for VABA, in the form of:

$$VABA_{NL} = a_0 + a_1 \times PV_{NL} + \varepsilon \tag{15}$$

The results of testing model (15) are presented in Table 4.9; given the values of $F = 11,399 > F_{0.05;1;9} = 5, 12$, the null hypothesis H_{0_5} was rejected and the alternative hypothesis H_{1_5} was accepted, and hence model (15) was statistically significant.

The $VABA_{NL}$ model describes the functional link between VABA and the value of the PV in the Netherlands in the period 2006–2015. For the significance chosen, the parameter a_1 was statistically significant (sign(LB) = sign (UB)) and the regression was of the form:

$$VABA_{NL} = 99.804 + 0.743 \times PV_{NL} + \varepsilon \tag{16}$$

Model (16), together with the CI of the regressor (a_1), indicate that with a €1 million rise in the value of PV, VABA in the Netherlands would increase by a value of between €0.235 million and €1249 million, with the most likely increase being €0.743 million.

The UK was the fourth state to be included in cluster C2 in 2015, based on VABA, PA, PV and TLF values.

Table 4.9. Results of model (15) testing, parameter values, test results and their CIs.

Model	F	Sig. F	R^2	Name	B	t	Sig.	LB	UB
								95% CI for B	
$VABA_{NL}$	11.399	0.009	0.587	a_0	99.804	0.04	0.972	−6163.73	6363.34
				a_1	0.743	3.37	0.009	0.235	1.249

Table 4.10. Characteristics of the mean values of the evolution of VABA, PA, PV and TLF in the UK, 2006–2015.

Variable	Mean	SD	SE	$t\_statistic$	LB	UB	V (%)
					95% CI for mean		
VABA	9,128.30	1,739.74	550.16	16.59	7,883.85	10,372.75	19.06
PA	14,044.65	2,179.72	689.29	20.38	12,485.47	15,603.82	15.52
PV	9,745.88	1,374.33	434.60	22.42	8,762.81	10,728.95	14.10
TLF	292.54	4.43	1.40	208.63	289.37	295.71	1.52

One of the first features was given by the mean of the values of the four variables recorded between 2006 and 2015 (Table 4.10). For this particular country also, for all the mean values, $t\_statistic > t_{\frac{\alpha}{2};9} = 2.262$, which led to rejection of the null hypothesis H_{0_2} and acceptance of the alternative hypothesis H_{1_2}, all the mean values being statistically significant.

The first difference from the mean values in Spain and the Netherlands was seen by the significantly higher values of V for VABA, PA and PV, which, although providing a good representability of the respective mean values, were determined by a larger scattering of the values of the variables around their mean values and implicitly obtaining relatively higher values of the SE.

Of the four variables, TLF recorded very small values of deviations from the mean, which resulted in a V value of 1.52%, indicating a very good representability of its mean value.

The evolutions of the values of the variables analysed during the period 2006–2015 are presented in Fig. 4.4, where, as in the case of the other countries, TLF is shown as TLF_h for a better graphical comparison.

It can be seen that, compared with Spain and the Netherlands, the trends for VABA, PA and PV values recorded for 2006–2015 in the UK had much higher slopes.

For example, whereas in the Netherlands the annual average increase in PA was €172.94 million, in the UK the increase was €363.97 million. However, the most relevant difference was between the annual average VABA of only €101.92 million in the Netherlands compared with €509.62 million in the UK.

In the agri-food sector in the UK the effects of the economic crisis have also been felt. Thus, in 2009, compared with 2008, the highest impact was recorded on the value of PV (−20.59%), followed by the value of VABA (−12.37%) and PA (−7.70%).

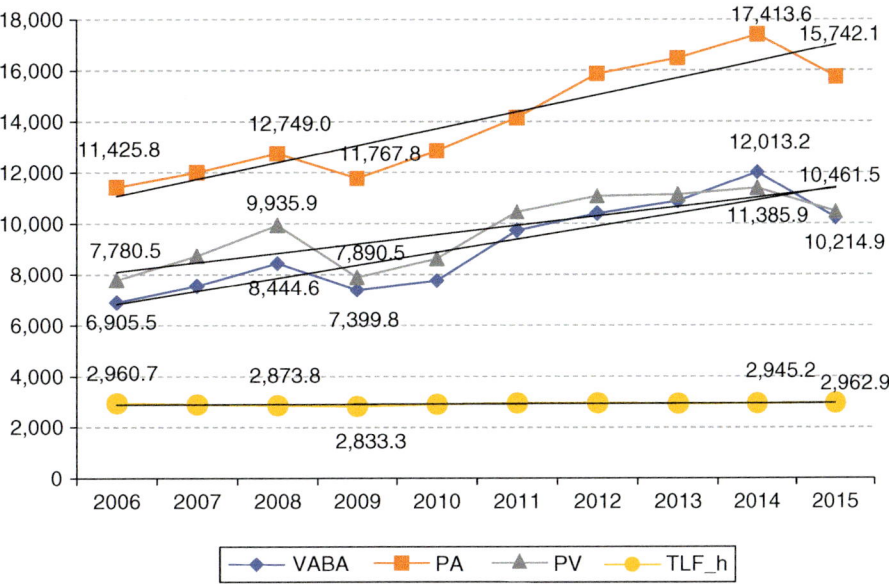

Fig. 4.4. Evolution of variables VABA, PA, PV and TLF_h, in the UK, 2006–2015. See Fig. 4.1 for *y*-axis units. (Based on information from Eurostat, 2017a,b.)

Another feature is that TLF in the UK over the period under review was approximately constant at around 296 (in 1000 AWUs). The intensity of the links between the variables analysed was indicated by the values of the Pearson correlation coefficients (Table 4.11). Apart from the TLF variable for which all values of the correlation coefficients were not statistically significant, between each of the three other variables the correlation coefficients were statistically significant for a 99% CI ($\alpha = 0.01$) and indicated very strong direct connections between them.

Given the values of the correlation coefficients in Table 4.11 in the case of the UK, for VABA the model proposed and tested was:

$$VABA_{GB} = a_0 + a_1 \times PA_{GB} + a_2 \times PV_{GB} + \varepsilon \tag{17}$$

The results of model (17) testing are presented in Table 4.12; as $F = 203.534 > F_{0.05;1;9} = 5.12$, the null hypothesis H_{0_5} was rejected and the alternative hypothesis H_{1_5} was accepted, and consequently model (17) was statistically significant.

The a_0 and a_1 parameters of model (17) were statistically significant for a 99% CI ($\alpha = 0.01$), and the parameter a_2 was significant for a 90% CI ($\alpha = 0.10$). Under these circumstances, the model realistically described the correlation between the indicators and showed that, for an increase of PA by €1 million while PV remained constant, VABA would increase by a value of between €0.365 million and €0.774 million. In contrast, an increase of PV by €1 million while PA remained constant would lead to an increase of VABA by a value of between €0.046 million and €0.694 million. Given that the

Table 4.11. Values of the correlation coefficients between VABA, PA, PV and TLF in the UK, 2006–2015, and their significance thresholds.

Variable		VABA	PA	PV	TLF
VABA	Pearson correlation	1	0.986**	0.957**	0.515
	Sig. (two-tailed)		0.000	0.000	0.127
PA	Pearson correlation	0.986**	1	0.932**	0.550
	Sig. (two-tailed)	0.000		0.000	0.100
PV	Pearson correlation	0.957**	0.932**	1	0.513
	Sig. (two-tailed)	0.000	0.000		0.129
TLF	Pearson correlation	0.515	0.550	0.513	1
	Sig. (two-tailed)	0.127	0.100	0.129	

*Correlation is significant at the 0.05 level (two-tailed).
**Correlation is significant at the 0.01 level (two-tailed).

Table 4.12. Results of model (17) testing, parameter values, test results and their CIs.

Model	F	Sig.F	R^2	Coefficients				90% CI interval for B	
				Name	B	t	Sig.	LB	UB
$VABA_{NL}$	203.534	0.000	0.983	a_0	−2475.87	−4.39	0.004	−3637.04	−1314.71
				a_1	0.569	5.27	0.001	0.365	0.774
				a_2	0.369	2.16	0.067	0.046	0.694

two factorial variables (PA and VA) would change simultaneously in the same direction (taking into account the trends of the two variables, the most likely change is in the direction of growth), then VABA in the UK would increase by a value of between €0.411 million and €1.468 million.

4.3   Characteristics of the Agri-food Sector in France and Italy

Cluster C3 comprised France and Italy in 2015, but compared with the other clusters, it showed the highest values of VABA (€30,790.6 million), PA (€20,504.3 million) and PV (€36,558.3 million).

In France, for the period 2006–2015, the characteristics of the mean values of the variables analysed are shown in Table 4.13. For all the mean values, $t\_statistic > t_{\frac{\alpha}{2};9} = 2.262$, which led to the rejection of the null hypothesis H_{0_2} and the acceptance of the alternative hypothesis H_{1_2}, all mean values being statistically significant. The values of V revealed that the mean values of the variables analysed had a very good representability. The evolutions of the mean values of VABA, PA, PV and TLF_h (expressed as 100 AWUs) are represented in Fig. 4.5.

In France, unlike in Germany and Spain, the slopes of the trends of the variables analysed were relatively similar. The greatest slope, corresponding to an annual average increase of €932.84 million, corresponded to the

Table 4.13. Characteristics of the mean values of the evolution of VABA, PA, PV and TLF in France, 2006–2015.

Variable	Mean	SD	SE	t_statistic	95% CI for mean		V (%)
					LB	UB	
VABA	25,960.71	2,876.26	909.55	28.54	23,903.30	28,018.12	11.08
PA	23,948.74	1,774.70	561.21	42.67	22,679.29	25,218.20	7.41
PV	39,524.31	3,401.37	1,075.61	36.75	37,091.29	41,957.33	8.61
TLF	814.59	41.20	13.03	62.53	785.12	844.06	5.06

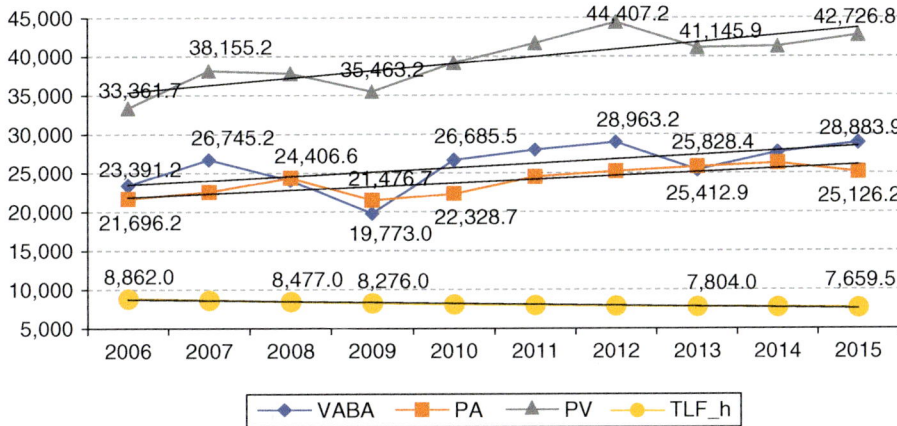

Fig. 4.5. Evolution of variables VABA, PA, PV and TLF_h, in France, 2006–2015. See Fig. 4.1 for *y*-axis units. (Based on information from Eurostat, 2017a,b.)

evolution of the value of PV, followed by VABA (€553.76 million) and PA (€470.09 million). At the same time, TLF showed an average reduction of 13.32 (in 1000 AWUs).

The economic crisis influenced the evolution of the French agrarian industry to a greater extent than in Spain but to a lesser extent than in Germany. If we refer only to VABA, this indicator decreased in 2009 compared with 2008 by 17.92% compared with only 9.41% in Spain and 22.55% in Germany.

In terms of TLF, the decrease recorded in the same period in France (−2.37%) was comparable to that in Germany (−2.01%) but in absolute terms was much lower than in Spain (−8.93%).

The values of the bilateral correlation coefficients between the variables are shown in Table 4.14. Except for the correlation coefficient between VABA and TLF (−0.555), which revealed a rather poor inverse correlation between these two variables and which was statistically significant only for a significance level of α =0.01, the bilateral correlation coefficients of the other variables were statistically significant for the significance level chosen (α = 0.05).

Table 4.14. Values of the correlation coefficients between VABA, PA, PV and TLF in France, 2006–2015, and their significance thresholds.

		VABA	PA	PV	TLF
VABA	Pearson correlation	1	0.641*	0.835**	−0.555
	Sig. (two-tailed)		0.046	0.003	0.096
PA	Pearson correlation	0.641*	1	0.813**	−0.762*
	Sig. (two-tailed)	0.046		0.004	0.010
PV	Pearson correlation	0.835**	0.813**	1	−0.845**
	Sig. (two-tailed)	0.003	0.004		0.002
TLF	Pearson correlation	−0.555	−0.762*	−0.845**	1
	Sig. (two-tailed)	0.096	0.010	0.002	

*Correlation is significant at the 0.05 level (two-tailed).
**Correlation is significant at the 0.01 level (two-tailed).

Table 4.15. Results of the testing of the model in Eqns 18, parameter values, test results and their CIs.

Model	F	Sig. F	R^2	Name	Coefficients B	t	Sig.	95% CI for B LB	UB
$VABA_{F1}$	8.195	0.015	0.701	a_{10}	−623.16	−0.07	0.94	−19,711.01	18,464.68
				a_{11}	0.783	2.61	0.04	0.072	1.494
				a_{12}	−0.182	−0.32	0.76	−1.544	1.180
$VABA_{F2}$	5.569	0.046	0.410	a_{20}	1,094.92	−0.11	0.92	−23,263.12	25,452.96
				a_{21}	1.038	2.36	0.04	0.024	2.053
$VABA_{F3}$	18.534	0.003	0.696	a_{30}	−1,931.43	0.29	0.76	−16,994.56	13,131.69
				a_{31}	0.706	4.28	0.00	0.362	1.086

As the values of the correlation coefficients between VABA and PA (0.641) and between VABA and PV (0.835) were statistically significant for $\alpha = 0.05$ and $\alpha = 0.01$, respectively, three models were tested for the case of France:

$$VABA_{F1} = a_{10} + a_{11} \times PA_F + a_{12} \cdot PV_F + \varepsilon_1$$
$$VABA_{F2} = a_{20} + a_{21} \times PA_F + \varepsilon_2 \tag{18}$$
$$VABA_{F3} = a_{30} + a_{31} \times PV_F + \varepsilon_3$$

The results obtained for models in Eqns 18 are presented in Table 4.15. From the three models tested for the $VABA_{F1}$ model, due to the fact that $F = 8.195 > F_{0.05;1;9} = 5.12$, the null hypothesis H_{0_5} was rejected and the alternative hypothesis was accepted. Consequently, the $VABA_{F1}$ model was statistically significant. The same conclusion was also reached by taking into account that Sig. $F = 0.015 < 0.05$.

However, because the null hypothesis H_{0_6} was accepted for parameter a_{12}, this proved that parameter a_{12} was not statistically significant for the chosen significance level and, consequently, the $VABA_{F1}$ model could not be used.

For the $VABA_{F2}$ and $VABA_{F3}$ models, the null hypothesis H_{0_5} was also rejected and the alternative hypothesis H_{1_5} was accepted, both models being statistically significant.

The $VABA_{F2}$ model describes the functional link between VABA and the value of PA in France over the period 2006–2015.

For the chosen significance level ($\alpha = 0.05$) for parameter a_{21}, sign(LB) = sign (UB), which led to rejection of the null hypothesis H_{0_6} and acceptance of the alternative hypothesis H_{1_6}. In conclusion, this model was statistically significant, and the regression model had the form:

$$VABA_{F2} = 1094.92 + 1.038 \times PA_F + \varepsilon_2 \tag{19}$$

Model (19), together with the CI of the regressor (a_{21}), indicated that the increase in the value of PA by €1 million would lead to an increase in VABA in France by a value of between €0.024 million and €2.053 million, the most likely increase being €1.038 million.

The $VABA_{F3}$ model describes the functional link between VABA and the value of PV. Parameter a_{31} was statistically significant (Sign(LB) = Sign(UB)), and the regression model had the form:

$$VABA_{F3} = -1931.43 + 0.706 \times PV_F + \varepsilon_3 \tag{20}$$

Model (20), together with the CI of the regressor (a_{31}) demonstrated that the increase in the value of PV by €1 million would lead to an increase in VABA in France by an amount between €0.362 million and €0.86 million, with the most likely increase being €0.607 million.

For Italy, the mean values of VABA, PA, PV and TLF over the analysed period are shown in Table 4.16. For all mean values, $t\_statistic > t_{\frac{\alpha}{2}\cdot9} = 2.262$, which led to rejection of the null hypothesis H_{0_2} and acceptance of the alternative hypothesis H_{1_2}.

In conclusion, all mean values were statistically significant for the significance level chosen ($\alpha = 0.05$). In addition, the values of the V showed that the mean values had a very good representability for all four variables.

In the case of Italy, the evolution of the values of the variables analysed during the period 2006–2015 highlighted the rather strong link between VABA and the value of PV (Fig. 4.6). This aspect was also evidenced by the value of the correlation coefficient between the two variables (0.926) presented in Table 4.17.

Table 4.16. Characteristics of the mean values of the evolution of VABA, PA, PV and TLF in Italy, 2006–2015.

Variable	Mean	SD	SE	t_statistic	95% CI for mean		V (%)
					LB	UB	
VABA	28,919.45	2,482.97	785.19	36.83	27,143.36	30,695.54	8.59
PA	15,403.99	1,153.33	364.72	42.24	14,579.00	16,228.98	7.49
PV	28,499.31	1,681.31	531.68	53.60	27,296.65	29,701.96	5.90
TLF	1,155.36	50.86	16.08	71.83	1,118.98	1,191.74	4.40

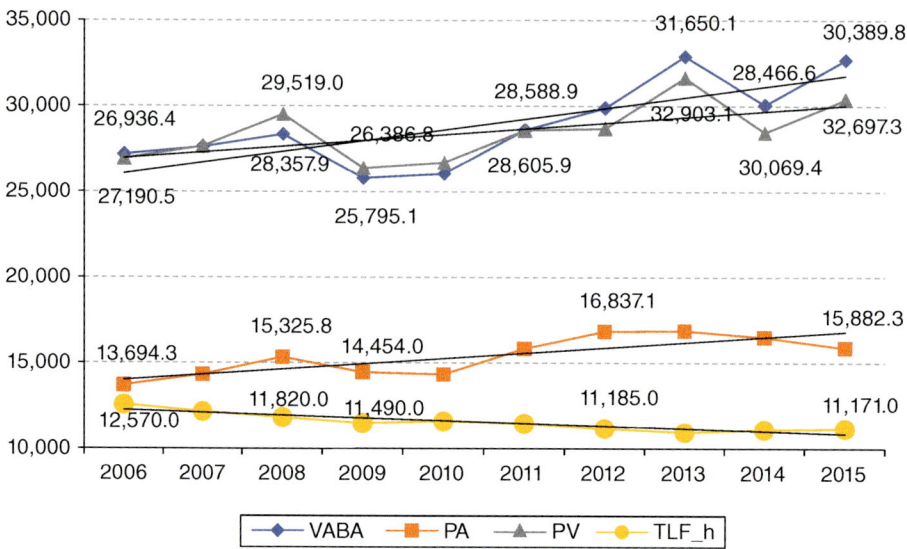

Fig. 4.6. Evolution of variables VABA, PA, PV and TLF_h, in Italy, 2006–2015. See Fig. 4.1 for *y*-axis units. (Based on information from Eurostat, 2017a,b.)

Table 4.17. Values of the correlation coefficients between VABA, PA, PV and TLF in Italy, 2006–2015, and their significance thresholds.

		VABA	PA	PV	TLF
VABA	Pearson correlation	1	0.800**	0.926**	−0.645*
	Sig. (two-tailed)		0.005	0.000	0.044
PA	Pearson correlation	0.800**	1	0.760*	−0.858**
	Sig. (two-tailed)	0.005		0.011	0.001
PV	Pearson correlation	0.926**	0.760*	1	−0.580
	Sig. (two-tailed)	0.000	0.011		0.079
TLF	Pearson correlation	−0.645*	−0.858**	−0.580	1
	Sig. (two-tailed)	0.044	0.001	0.079	

*Correlation is significant at the 0.05 level (two-tailed).
**Correlation is significant at the 0.01 level (two-tailed).

The evolution of the values of the analysed variables show significant annual increases in VABA, the annual average increase being €631.82 million. The values of PA and PV recorded similar yearly increases of €310.90 million and €340.67 million, respectively, while TLF decreased by 15.37 (in 1000 AWUs).

The economic crisis also influenced the evolution of the food industry in Italy. Compared with 2008, the impact on VABA (−9.04%) and PV (−10.62%) was similar to that in Germany, while the impact on PA (−5.69%) was similar to that in Spain. The impact on TLF (−2.79%) was similar to that in Germany (−2.01%) and Spain (−2.37%).

The values of the Pearson bilateral parametric correlation coefficients between the variables analysed are presented in Table 4.17. Except for the correlation coefficient between PV and TLF (−0.580), which was statistically significant only for a significance level of $\alpha = 0.01$, the bilateral correlation coefficients between the other variables were statistically significant for the significance level chosen ($\alpha = 0.05$).

As the values of the correlation coefficients between VABA and PA (0.800), VABA and PV (0.926), and VABA and TLF (−0.645) were statistically significant for $\alpha = 0.05$, in the case of Italy four models were tested:

$$VABA_{I1} = a_{10} + a_{11} \times PA_I + a_{12} \times PV_I + a_{13} \times TLF_I + \varepsilon_1$$
$$VABA_{I2} = a_{20} + a_{21} \times PA_I + \varepsilon_2$$
$$VABA_{I3} = a_{30} + a_{31} \times PV_I + \varepsilon_3 \qquad (21)$$
$$VABA_{I4} = a_{40} + a_{41} \times TLF_I + \varepsilon_4$$

The results obtained for the models in Eqns 21 are presented in Table 4.18. Of the four models tested, for the $VABA_{I1}$ model, as $F = 14.759 > F_{0.05;1;9} = 5.12$, the null hypothesis H_{0_5} was rejected and the alternative hypothesis was accepted. Consequently, the $VABA_{I1}$ model was statistically significant. However, because for parameters a_{11} and a_{13} the null hypothesis H_{0_6} was accepted, parameters a_{11} and a_{13} were not statistically significant for the chosen significance level and subsequently the $VABA_{I1}$ model could not be used.

Concerning the other three models, the null hypothesis H_{0_5} was also rejected and the alternative hypothesis H_{1_5} was accepted; thus, models $VABA_{I2}$, $VABA_{I3}$ and $VABA_{I4}$ were statistically significant.

Model $VABA_{I2}$ describes the functional connection between VABA and the value of PA in Italy over the period 2006–2015. For the chosen significance level ($\alpha = 0.05$), for parameter a_{21}, sign(LB) = sign (UB), leading to the rejection of the null hypothesis H_{0_6} and acceptance of the alternative hypothesis H_{1_6}. The regression model therefore becomes:

$$VABA_{I2} = 2374.95 + 1.72 \times PA_I + \varepsilon_2 \qquad (22)$$

Table 4.18. Results of the model (21) testing, parameter values, test results and their CIs.

Model	F	Sig. F	R^2	Name	B	t	Sig.	LB	UB
								95% CI for B	
					Coefficients				
$VABA_{I1}$	14.759	0.004	0.881	a_{10}	−6,315.46	−0.26	0.80	−64,641.48	52,010.55
				a_{11}	0.39	0.51	0.63	−1.47	2.25
				a_{12}	1.12	3.42	0.01	0.32	1.93
				a_{13}	−2.39	−0.17	0.87	−36.02	31.24
$VABA_{I2}$	14.265	0.005	0.641	a_{20}	2,374.95	0.34	0.75	−13,872.72	18,622.62
				a_{21}	1.72	3.78	0.01	0.67	2.78
$VABA_{I3}$	48.327	0.000	0.858	a_{30}	−10,065.3	−1.79	0.11	−23,017.3	2,886.73
				a_{31}	1.368	6.95	0.00	0.914	1.822
$VABA_{I4}$	5.704	0.044	0.416	a_{40}	65,308.4	4.28	0.00	30,143.47	10,0473.3
				a_{41}	−31.495	−2.38	0.04	−61.905	−1.08593

Taking into account the limits of the CI, according to model (22), the increase in value of PA by €1 million would lead to an increase in VABA in Italy ranging from €0.67 million to €2.78 million, with the most probable increase being €1.72 million.

Model $VABA_{I3}$ describes the functional link between VABA and the value of PV. Parameter a_{31} was statistically significant (sign(LB) = sign (UB)), and the corresponding regression model was:

$$VABA_{F3} = -10065.3 + 1.368 \times PV_F + \varepsilon_3 \tag{23}$$

According to model (23) for a 95% CI, the increase by €1 million in the value of PV in Italy would lead to an increase in VABA ranging between €0.914 million and €1.822 million, with the most likely increase being €1.368 million.

With regard to the $VABA_{I4}$ model, although both the model itself and parameter a_{41} were statistically significant for the significance level chosen ($\alpha = 0.05$), taking into account the fact that the value of the determinant coefficient (R^2) is less than 0.5 (the share of TLF influence on VABA is less than 50%, with other factors being predominant), we did not consider it conclusive for the analysis.

4.4   Progression Structure of the Agrarian Sector in Poland and Romania

An important feature of cluster C4, comprising Poland and Romania, was the very high value of the total population involved in the agri-food sector, with the mean TLF value for this cluster being 2410 (in 1000 AWUs).

The characteristics of the mean values of VABA, PA, PV and TLF recorded in Poland during 2006–2015 are presented in Table 4.19. Given that for all the mean values the statistic test (*t_statistic*) was much higher than the corresponding critical value (2.262), the null hypothesis H_{0_2} was rejected and the alternative hypothesis H_{1_2} was accepted, all mean values being statistically significant.

Considering the values of *V*, it could be concluded that the mean values for TLF, PA and PV showed very good representability, while for VABA there was a good representability.

The evolution of the variables analysed in the period 2006–2015 in Poland is illustrated in Fig. 4.7. Unlike the evolution of VABA, PA and PV in clusters C2

Table 4.19. Characteristics of the mean values of the evolution of VABA, PA, PV and TLF in Poland, 2006–2015.

| Variable | Mean | SD | SE | *t_statistic* | 95% CI for mean | | *V* (%) |
					LB	UB	
VABA	7,157.79	1,171.07	370.32	19.33	6,320.12	7,995.46	16.36
PA	9,671.22	1,138.90	360.15	26.85	8,856.55	10,485.88	11.78
PV	10,692.22	1,463.45	462.78	23.10	9,645.41	11,739.04	13.69
TLF	2,066.00	182.62	57.75	35.77	1,935.37	2,196.63	8.84

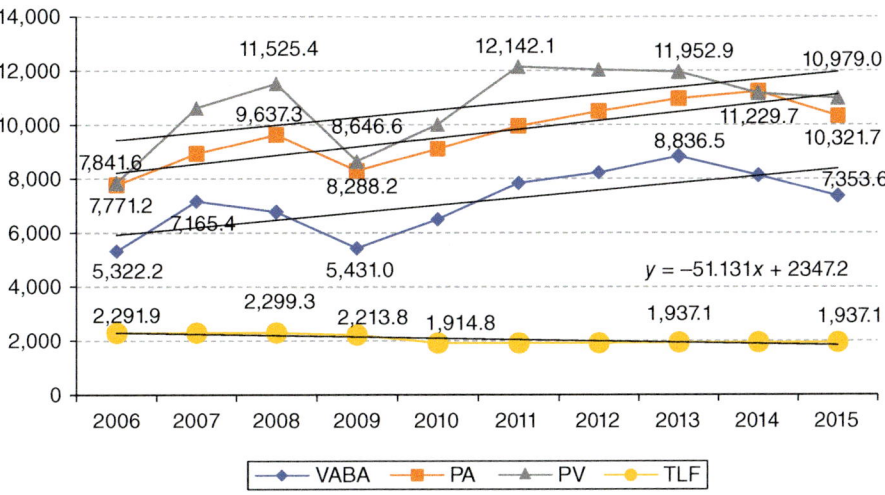

Fig. 4.7. Evolution of variables VABA, PA, PV and TLF, in Poland, 2006–2015. See Fig. 4.1 for *y*-axis units. (Based on information from Eurostat, 2017a,b.)

and C3 described previously, in the case of Poland, the impact of the economic crisis was most evident.

Thus, as a result of the economic crisis in Poland in 2009, compared with 2008, the value of PA decreased from €9637.3 million to €8288.2 million (−14.0%), the value of PV decreased from €11,525.4 million to €8646.6 million (−24.98%), and VABA decreased from €6775.3 million to €5431.0 million (−19.84%).

Since 2010, the agri-food sector has begun to recover, but from 2014 there has been a decline in VABA, PA and PV, with PA in 2015 being 8.1% lower than in the previous year and VABA recording a 9.55% reduction.

The values of the Pearson correlation coefficients and the corresponding significance thresholds are presented in Table 4.20. Except for the correlation coefficient between PV and TLF whose value was not statistically significant for the significance threshold chosen ($\alpha = 0.05$), the values of the other coefficients determined rejection of the null hypothesis H_{0_4} and acceptance of the alternative hypothesis H_{1_4}, the values of the correlation coefficients being statistically significant.

A significant characteristic of the evolution of the four variables was that, in the case of Poland, between VABA and the other three variables, the correlation coefficient values were statistically significant, indicating strong direct correlations between VABA and PA and PV, as well as an inverse correlation of medium intensity between VABA and TLF. Considering the values of the correlation coefficients, for the resulting variable VABA, the PA and PV determining variables were chosen. The tested model had the form:

$$\text{VABA}_{PL} = a_0 + a_1 \times \text{PA}_{PL} + a_2 \times \text{PV}_{PL} + \varepsilon \tag{24}$$

The results of testing model (24) are presented in Table 4.21. Given that $F = 35,732 > F_{0.05;1;9} = 5.12$, the null hypothesis H_{0_5} was rejected and the

Table 4.20. Values of the correlation coefficients between VABA, PA, PV and TLF in Poland, 2006–2015, and their significance thresholds.

		VABA	PA	PV	TLF
VABA	Pearson correlation	1	0.928**	0.904**	−0.670*
	Sig. (two-tailed)		0.000	0.000	0.034
PA	Pearson correlation	0.928**	1	0.848**	−0.721*
	Sig. (two-tailed)	0.000		0.002	0.019
PV	Pearson correlation	0.904**	0.848**	1	−0.560
	Sig. (two-tailed)	0.000	0.002		0.092
TLF	Pearson correlation	−0.670*	−0.721*	−0.560	1
	Sig. (two-tailed)	0.034	0.019	0.092	

*Correlation is significant at the 0.05 level (two-tailed).
**Correlation is significant at the 0.01 level (two-tailed).

Table 4.21. Results of the model (24) testing, parameter values, test results and their CIs.

Model	F	Sig. F	R^2	Coefficients				90% CI for B	
				Name	B	t	Sig.	LB	UB
$VABA_{NL}$	35.732	0.000	0.911	a_0	−2126.55	−1.88	0.101	−4267.3	14.199
				a_1	0.591	2.70	0.031	0.176	1.006
				a_2	0.334	1.96	0.091	0.011	0.656

alternative hypothesis H_{1_5} was accepted, and consequently the $VABA_{PL}$ model was statistically significant. Parameter a_1 of the model was statistically significant for a 95% CI ($\alpha = 0.05$) and parameter a_2 for a 90% CI ($\alpha = 0.10$). The model in this case was:

$$VABA_{PL} = -2126.55 + 0.591 \times PA_{PL} + 0.334 \times PV_{PL} + \varepsilon \qquad (25)$$

In these circumstances, for the analysis carried out, model (25) describes the correlation between these indicators and indicated that at an increase of PA by €1 million, with a constant PV, would lead to an increase in VABA by a value between €0.176 million and €1.006 million. In contrast, for an increase in PV of €1 million, with PA remaining constant, VABA would increase by an amount between €0.011 million and €0.656 million.

Given that the two factorial variables (PA and VA) would change simultaneously in the same direction (taking into account the trends of the two variables, the most likely change is in the direction of growth), then VABA in Poland would change with a value between €0.187 million and €1.662 million.

For Romania, the second country in cluster C4, the mean values of VABA, PA, PV and TLF recorded between 2006 and 2015 are presented in Table 4.22. Considering that for all mean values the test statistic (*t_statistic*) was much higher than the corresponding critical value (2.262) the null hypothesis H_{0_2} is rejected and the alternative hypothesis H_{1_2} is once again accepted, all mean values being statistically significant.

The values of the coefficients of variation (Table 4.22) indicate a very good representability of the mean value for PA and VABA and good representability for PV and TLF. These differences were due to the varying degrees of scattering of the values of the variables around their mean values, which are more reduced for the PA variable and relatively extensive for TLF.

The evolution of the variables analysed in the period 2006–2015 in Romania is illustrated in Fig. 4.8. The evolutions of PV, VABA, and TLF oscillated significantly around the trends, reinforcing the conclusions on the scattering of values around the mean for these variables. If the mean annual increase in PV was not very large, the difference was significant for PA. In Poland, the annual increase in PA in the analysed period was €322.19 million, compared with Romania where the mean annual growth was negative (–€5.02 million).

These differences also spread to VABA, which in Poland recorded an average annual increase of €273.17 million, while in Romania it was only €16.29 million. However, the size of TLF in Romania registered an annual average decrease of –127.25 (in 1000 AWUs) compared with only –51.13 in Poland.

Table 4.22. Characteristics of the mean values of the evolution of VABA, PA, PV and TLF in Romania, 2006–2015.

Variable	Mean	SD	SE	t_statistic	95% CI for mean LB	95% CI for mean UB	V (%)
VABA	6,826.47	811.29	256.55	26.61	6,246.15	7,406.79	11.88
PA	3,903.92	134.52	42.54	91.78	3,807.70	4,000.14	3.45
PV	10,278.88	1,706.31	539.58	19.05	9,058.35	11,499.42	16.60
TLF	1,810.30	408.74	129.26	14.01	1,517.92	2,102.68	22.58

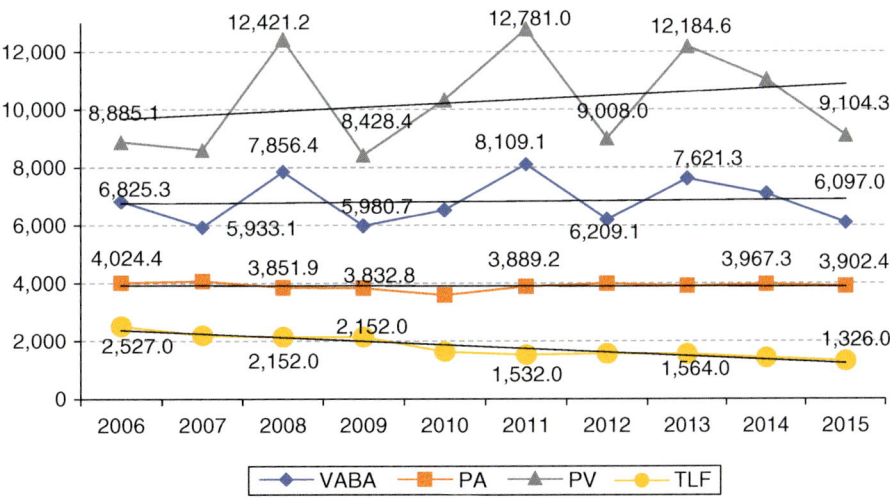

Fig. 4.8. Evolution of variables VABA, PA, PV and TLF_h, in Romania, 2006–2015. See Fig. 4.1 for *y*-axis units. (Based on information from Eurostat, 2017a,b.)

The values of the Pearson correlation coefficients as well as the significance thresholds corresponding to the values of the variables analysed in Romania are presented in Table 4.23. Except for the correlation coefficient between VABA and PV whose value was statistically significant for the significance threshold chosen ($\alpha = 0.05$), for all other coefficients the null hypothesis H_{0_4} was accepted, and hence the values of the correlation coefficients were not statistically significant.

Taking into account the fact that only the correlation between VABA and PV was statistically significant, for the evolution of VABA, a single model was tested in the case of Romania:

$$VABA_{RO} = a_0 + a_1 \times PV_{RO} + \varepsilon \qquad (26)$$

The results of the tests conducted on model (26) are shown in Table 4.24. As $F = 72{,}242 > F_{0.05;1;9} = 5.12$, the null hypothesis H_{0_5} was rejected and the alternative hypothesis H_{1_5} was accepted and consequently the model $VABA_{RO}$ was statistically significant.

The $VABA_{RO}$ model describes the functional link between VABA and the value of PV in Romania in the period 2006–2015. For the significance level chosen, the parameters a_0 and a_1 were statistically significant (sign(LB) = sign (UB)). The regression model became:

$$VABA_{RO} = 2189.265 + 0.451 \times PV_{Ro} + \varepsilon \qquad (27)$$

Model (27), together with the CI of the regressor (a_1), indicated that the increase in the value of PV by €1 million would lead to an increase in VABA

Table 4.23. Values of the correlation coefficients between VABA, PA, PV and TLF in Romania, 2006–2015, and their significance thresholds.

		VABA	PA	PV	TLF
VABA	Pearson correlation	1	−0.114	0.949**	−0.133
	Sig. (two-tailed)		0.755	0.000	0.715
PA	Pearson correlation	−0.114	1	−0.265	−0.255
	Sig. (two-tailed)	0.755		0.459	0.478
PV	Pearson correlation	0.949**	−0.265	1	−0.342
	Sig. (two-tailed)	0.000	0.459		0.333
TLF	Pearson correlation	−0.133	−0.255	−0.342	1
	Sig. (two-tailed)	0.715	0.478	0.333	

**Correlation is significant at the 0.01 level (two-tailed).

Table 4.24. Results of model (26) testing, parameter values, test results and their CIs.

Model	F	Sig. F	R^2	Name	B	t	Sig.	LB	UB
								95% CI for B	
					Coefficients				
$VABA_{RO}$	72.242	0.000	0.901	a_0	2189.26	3.96	0.004	915.64	3462.89
				a_1	0.451	8.49	0.000	0.329	0.574

by a value ranging between €0.329 million and €0.574 million, the most likely increase being €0.451 million.

Accordingly, the role and influence of the cultural model in displaying the agricultural production structures in the context of the transformation of the European agricultural model, implicitly on the level of productivity and the evolution of the rural communities, has been argued. From the analysis conducted throughout this study, it can be concluded that, inherently, the evolution of the CAP can be considered a result of the European sectoral cultural influences and experiences. From the analysis of some of the most representative performance indicators of agricultural production structures in some of the EU-28 Member States, the significant differences and dispersion of performance levels between Member States according to the characteristics of their specific cultural model is also supported.

Note

[1] This chapter is based on research carried by Andrei (2016).

5 Conclusions

The role of the CAP at the European level is complex and has various influences on the economies of the Member States, being an essential policy in the functioning of the EU. The CAP provides the functionality of a strategic economic sector through which both European food production and the supply of food products and feed needs are achieved. Given the multifunctional nature of agriculture, the CAP also contributes to the intrinsic development of rural communities, emphasizing the economic, social and environmental development elements among these communities and, last but not least, supporting farmers' incomes while reducing rural economic gaps and social disparities.

Starting from the fact that the CAP is the first and oldest of the European common policies, it can be seen that there have been many changes of paradigm in its functioning, generated by the many reforms imposed over time in the agricultural sector and on the agricultural economy of the EU.

Romania's integration into the EU has highlighted the existing differences between the national agricultural sector, which is at an early stage in the potential identification and recovery of rural economic gaps, and the objective need to comply with the requirements and practices of the European agricultural model. Over the years, agricultural sector-specific production relationships have changed; sectoral imbalances have partially diminished, leading to a degree of convergence with the CAP. Although the turning point in the field has changed, to the partial advantage of the Romanian agricultural sector, there is an objective need to capitalize on the agricultural potential, which has not yet materialized in the competitiveness parameters and hence is still generating imbalances.

The persistence of the family farm model in Europe, which is also well represented in Romania's agriculture, is essential in promoting a much more environmentally responsible agriculture and has a positive effect on job creation in rural areas and communities, which face a significant ageing process of the population. Restoring the generations of families to the farm, as well as attracting new farmers to support agricultural practices with a high level of productivity, may represent a new way for Romania's agriculture to capitalize on the agricultural potential. Through CAP reforms, combinations of measures have been introduced and promoted to improve the role and position of the agricultural sector in the national economic structure. Although the influence of the CAP on the development of the national agricultural sectors of the Member States is a determining factor, which has produced defining transformations and significant influences on European agriculture, there is still a need for a rethinking of national policies in the field and, above all, for the diversification of internal support mechanisms and resource orientation.

For Romania, not only is agriculture a fundamental economic sector with multiple functions and roles in the economic composition, with an appreciable development potential and providing income for a significant part of the rural population, it also provides a safety net, especially for rural areas, confronted with massive development inequalities. Sustainable development of the national agricultural sector, based on productivity and competitiveness, will help to improve the conditions in rural areas and strengthen national economic growth. The dynamics of the national agricultural sector have shown that it can become an economic branch with a vast potential for growth, diminishing the regional imbalances.

The available resources are limited, and there is only modest growth potential for family agricultural farms producing products in small farms and farmhouses mainly for subsistence or semi-subsistence, and the lack of an adequate rural infrastructure and organized markets, which are almost non-existent, accentuate these disadvantages. Moreover, considering other aspects, such as the level of education in this sector, the lack of an adequate fiscal system and the reduced desire of farmers to associate in groups, result in a picture with worrying nuances for Romanian agriculture.

Evolution of the Romanian economy and its transformation into a functional and highly competitive market economy cannot be achieved without proper and long-term adaptation of the national agricultural sector. Progressing through the centralized system of economic organization to a liberal system with a functional market economy has imposed a wide process of reform and adaptation to the new demands, which unfortunately still require a broad approach. The national agricultural economy still faces low levels of productivity, with poor capital endowment of agricultural holdings, but especially due to the polarization of farms. From our investigations and analysis in this book, we have noted the often antagonistic tendencies of the economic processes and phenomena specific to the agricultural economy, starting from the concentration of added value created in rural areas around agriculture and the employment of rural labour predominantly in this sector, but especially the low significant convergence compared with the European agricultural model and the poor attraction of funding to rural areas. Achieving a competitive agriculture that meets the demands of the European agricultural model requires re-establishment of the national agricultural policy focused on stimulating sustainable rural development against the background of improving the technical and exploitation capacity of agricultural farms and endowing small peasant farms with the necessary capital. As stated previously, for Romania, and especially for rural areas, agriculture continues to represent the safety net of the rural population, who are faced with low income levels and little possibility of diversification.

Modernization of the national agriculture and its placement on the market economy principles have implied, as derived from the present study, both sustainable growth of production and its orientation towards the market, as well as the development of rural areas and communities. The possibility of earning some income from agricultural activities has contributed, albeit to a limited extent, to this potential, leading to an improvement in the quality of life in rural areas and in farmers' incomes, providing some perspective for the large Romanian rural population.

Although, in the case of Romania, a significant proportion of the rural population, which is largely dependent on agricultural activities and subsistence farming, still plays a significant role in agriculture, being based on land ownership of small and fragmented plots, the developments resulting from the application of the CAP measures are encouraging and are proving to be effective, at least in terms of financial support for farms. The transition to a high-performance agriculture in Romania with high productivity and adaptability to meet the demands of the market economy implies not only convergence with the requirements of the CAP and the European agricultural model, but also the foundation and application of an integrated national agricultural policy in a broad bioeconomic strategy, in which food and food security policy represent the specific guidelines. Recent developments in European agricultural markets provide ample examples of potential imbalances that cannot be avoided by the national agricultural sector.

Diversifying national agriculture, putting it on a competitive and high efficiency base, as well as identifying innovative market management tools and financial support for farmers' activities, will provide opportunities for sectoral adjustment and convergence with the European agricultural model. Romanian agriculture, although part of the EU agriculture, will need to develop its own resistance to climate change and environmental exigencies, and at the same time respond to the shocks of the food supply base and the change of classical agricultural paradigms.

However, as shown by the analyses of EU countries in various years, we can see a repositioning of countries within clusters from one period to another, allowing the conclusion that every EU country is directly influenced by CAP measures and changes, as well as by European and international economic contexts. Based on these analyses, the new direction of the CAP post-2020 must acknowledge the two-way impact – not only of CAP actions upon the Member States but also the influence of the Member States on how the new CAP addresses agricultural and rural needs. Under these circumstances, vulnerable countries in terms of sustainable and efficient agricultural sectors, such as Romania, may need to accelerate the implementation of articulate economic and agricultural measures in order to keep up with the ever-shifting agrarian needs of EU citizens and worldwide consumers, in order to avoid widening gaps compared with other Member States.

References

Aceleanu, M.I., Molănescu, A.G., Crăciun, L. and Voicu, C. (2015) The status of Romanian agriculture and some measures to take. *Theoretical and Applied Economics* 22, 123–138.

Aggarwal, C.C. and Reddy, C.K. (2013) *Data Clustering: Algorithms and Applications.* CRC Press, Oxford.

Anderson, K. and Tyers, R. (1993) Implications of EC Expansion for European Agricultural Policies, Trade and Welfare. *CEPR Discussion Papers* 829, 209–237.

Andrei, J.V. (2016) Romanian agriculture paradigm's shifts in context of convergence to the European agricultural model. Habilitation thesis, Bucharest University of Economic Studies, Bucharest, Romania.

Andrei, J.V. and Begalli, D. (2015) Romania and the European economic and cultural model. Changing the paradigms. *Procedia Economics and Finance* 22, 370–379.

Andrei, J.V. and Dușmănescu, D. (2012) Some possible effects of common agricultural policy (CAP) on Romanian agriculture during the 2014–2020 financial perspective: how much does the paradigm really change? In: *3rd International Symposium on Agrarian Economy and Rural Development – Realities and Perspectives for Romania.* Research Institute for Agriculture Economy and Rural Development, Bucharest, Romania, pp. 8–14.

Andrei, J.V., Dușmănescu, D. and Mieilă, M. (2015) The influences of the cultural models on agricultural production structures in Romania and some EU-28 countries – a perspective. *Economics of Agriculture* 62, 1–17.

Andren, T. (2007) Econometrics. Available at: http://pusbindiklat.lipi.go.id/psb/files/original/c74bccf36d904dbc384f51a6a5d653bb.pdf (accessed 13 April 2018).

Anghelache, C. (2018) Structural analysis of Romanian agriculture. *Romanian Statistical Review Supplement* 66, 11–18.

Araya, I., Trombettoni, G., Neveu, B. and Chabert, G. (2014) Upper bounding in inner regions for global optimization under inequality constraints. *Journal of Global Optimization* 60, 145–164.

Bularca (Olaru), E. and Toma, E. (2018) Structural change in the Romanian agriculture: implications for the farming sector. *Scientific Papers Series Management, Economic Engineering in Agriculture and Rural Development* 18, 59–66.

Buller, H. and Hoggart, K. (2017) *Agricultural Transformation, Food and Environment.* Routledge, London.

Cardinal, R.N. and Aitken, M.R.F. (2013) *ANOVA for the Behavioural Sciences Researcher.* Lawrence Erlbaum Associates, Mahwah, New Jersey.

Cartwright, A.L. (2017) *The Return of the Peasant. Land Reform in Post-Communist Romania.* Routledge, London.

Ciampi Stančová, K. and Cavicchi, A. (2019) EU policies and instruments to support the agri-food sector. In: *Smart Specialisation and the Agri-food System.* Springer International Publishing, Cham, Switzerland, pp. 25–42.

Ciutacu, C., Chivu, L. and Andrei, J.V. (2015) Similarities and dissimilarities between the EU agricultural and rural development model and Romanian agriculture. Challenges and perspectives. *Land Use Policy* 44, 169–176.

Constantin, F. (2017) Study on the evolution of labor productivity in Romanian agriculture compare to some EU countries. *Quality – Access to Success* 18, 135–140.

Cornish, R. (2006). Oneway analysis of variance. Available at: http://www.lboro.ac.uk/media/wwwlboroacuk/content/mlsc/downloads/1.5_OnewayANOVA.pdf (accessed 11 April 2017).

Croitoru, G., Mieilă, M. and Cicea, C. (2018) The competitive position of the Romanian agriculture. *Economics of Agriculture* 57(Special nu), 161–169.

Daugbjerg, C. and Swinbank, A. (2007) The politics of CAP reform: trade negotiations, institutional settings and blame avoidance. *Journal of Common Market Studies* 45, 1–22.

Davidova, S., Fredriksson, L., Gorton, M., Mishev, P. and Petrovici, D. (2012) Subsistence farming, incomes, and agricultural livelihoods in the new Member States of the European Union. *Environment and Planning C: Government and Policy* 30, 209–227.

Dipti, M. and Patel, T. (2014) *K*-means based data stream clustering algorithm extended with no. of cluster estimation method. *International Journal of Advance Engineering and Research Development* 1. Available at: http://ijaerd.com/papers/finished_papers/ijaerd%2014-0243.pdf (accessed 15 March 2019).

Dougherty, C. (2007) *Introduction to Econometrics,* 3rd edn. Oxford University Press, Oxford.

Drăgan, G. (2005) *Uniunea Europeană între Federalism și Interguvernamentalism: Politici Comune ale UE.* ASE Publishing House, Bucharest, Romania.

Drăgoi, M.C. (2016) Health determinants: nutrition-related facts. In: Andrei, J.V. (Ed.) Food Science, *Production, and Engineering in Contemporary Economies.* IGI Global, Hershey, Pennsylvania, pp. 393–417

Drăgoi, M.C., Andrei, J.V., Mieilă, M., Panait, M., Dobrotă, C.E. and Lădaru, R.G. (2018) Food safety and security in Romania – an econometric analysis in the context of national agricultural paradigm transformation. *Amfiteatru Economic* 20, 134–150.

Drăgoi, M.C., Iamandi, I.-E., Munteanu, S.M., Ciobanu, R., Țarțavulea (Dieaconescu), R.I. and Lădaru, R.G. (2017) Incentives for developing resilient agritourism entrepreneurship in rural communities in Romania in a European context. *Sustainability* 9, 1–30.

EurActiv (2017) Plantând semințele PAC post-2020. Available at: http://www.euractiv.ro/agricultura/plantand-semintele-pac-post-2020-8377 (accessed 15 September 2017).

European Commission (2009) Health Check of the CAP (current situation, Commission proposal and Council outcome). Available at: https://ec.europa.eu/agriculture/sites/agriculture/files/policy-perspectives/impact-assessment/cap-health-check/documents/before_after_en.pdf (accessed 14 December 2017).

European Commission (2012) *The Common Agricultural Policy. A Story to be Continued.* Office for Official Publications of the European Communities, Luxembourg.

European Commission (2013) Overview of CAP reform 2014–2020. Available at: https://ec.europa.eu/agriculture/sites/agriculture/files/policy-perspectives/policy-briefs/05_en.pdf (accessed 21 July 2017).

European Commission (2014) Politica agricolă comună – aprofundarea PAC. Available at: https://ec.europa.eu/agriculture/cap-for-our-roots/cap-in-depth/index_ro.htm (accessed 12 October 2017).

European Commission (2017a) EU action for smart villages. Available at: https://ec.europa.eu/agriculture/sites/agriculture/files/rural-development-2014-2020/looking-ahead/rur-dev-small-villages_en.pdf (accessed 5 January 2018).

European Commission (2017b) Impact assessment for the 2008 CAP "Health Check". Available at: https://ec.europa.eu/agriculture/policy-perspectives/impact-assessment/cap-health-check_it (accessed 14 December 2017).

European Commission (2017c) Multiannual financial framework 2014–2020 and the financing of the CAP. Available at: https://ec.europa.eu/agriculture/sites/agriculture/files/cap-funding/budget/mff-2014-2020/mff-figures-and-cap en.pdf (accessed 3 October 2017).

European Commission (2017d) Sicco Mansholt: farmer, resistance fighter and a true European. Available at: https://europa.eu/european-union/sites/europaeu/files/docs/body/sicco_mansholt_en.pdf (accessed 12 July 2017).

European Commission (2017e) The future of food and farming. Agriculture 2.0. Available at: https://ec.europa.eu/agriculture/sites/agriculture/files/future-of-cap/factsheet_v_en.pdf (accessed 21 January 2018).

European Commission (2017f) The future of food and farming. 2017, Available at: https://ec.europa.eu/agriculture/sites/agriculture/files/future-of-cap/future_of_food_and_farming_communication_en.pdf (accessed 20 December 2017).

European Commission (2017g) The common agricultural policy at a glance. Available at: https://ec.europa.eu/agriculture/cap-overview/history_en (accessed 22 August 2017).

European Commission (2018a) Common agricultural policy. Available at: https://ec.europa.eu/info/food-farming-fisheries/key-policies/common-agricultural-policy_en (accessed 21 August 2018).

European Commission (2018b) Statistical factsheet: Romania. Available at: https://ec.europa.eu/agriculture/sites/agriculture/files/statistics/factsheets/pdf/ro_en.pdf (accessed 10 October 2018).

European Institute of Romania (2003) Politica agricolă. Available at: http://beta.ier.ro/documente/formare/Politica_agricola.pdf (accessed 11 November 2017).

European Network for Rural Development (2010) Semi-subsistence farming in Europe: concepts and key issues. Available at: https://enrd.ec.europa.eu/sites/enrd/files/fms/pdf/FB3C4513-AED5-E24F-E70A-F7EA236BBB5A.pdf (accessed 11 July 2017).

European Parliament (2017) Fișe tehnice privind Uniunea Europeană. Politica agricolă comună (PAC) și tratatul. Available at: http://www.europarl.europa.eu/factsheets/ro/sheet/103/politica-agricola-comuna-pac-si-tratatul (accessed 30 September 2017).

Eurostat (2015a) Farm structure statistics. Available at: http://ec.europa.eu/eurostat/statistics-explained/index.php/Farm_structure_statistics (accessed 5 September 2017).

Eurostat (2015b) Land cover, land use and landscape. Available at: https://ec.europa.eu/eurostat/statistics-explained/index.php?title=Land_use_statistics (accessed 1 October 2017).

Eurostat (2016a) Agricultural accounts and prices. Available at: https://ec.europa.eu/eurostat/statistics-explained/index.php?title=Archive:Agricultural_accounts_and_prices (accessed 15 March 2019).

Eurostat (2016b) Small and large farms in the EU – statistics from the farm structure survey. Available at: https://ec.europa.eu/eurostat/statistics-explained/index.php?title=Archive:Small_and_large_farms_in_the_EU_-_statistics_from_the_farm_structure_survey (accessed 15 March 2019).

Eurostat (2017a) Agricultural labour input statistics: absolute figures (1,000 annual work units). Available at: https://ec.europa.eu/eurostat/web/products-datasets/-/aact_ali01 (accessed 15 March 2019).

Eurostat (2017b) Economic accounts for agriculture – agricultural income (indicators A, B, C). Available at: http://ec.europa.eu/eurostat/web/products-datasets/product?code=aact_eaa06 (accessed 11 July 2017).

Feher, A., Goşa, V., Raicov, M., Haranguş, D. and Condea, B.V. (2017) Convergence of Romanian and Europe Union agriculture – evolution and prospective assessment. *Land Use Policy* 67, 670–678.

Feichtinger, P. and Salhofer, K. (2015) *The Fischler Reform of the Common Agricultural Policy and Agricultural Land Prices. Working Papers*. Institute for Sustainable Economic Development, Department of Economics and Social Sciences, University of Natural Resources and Life Sciences, Vienna, Austria.

Galluzzo, N. (2017) Medium term analysis of technical and allocative efficiency in Romanian farms using FADN dataset. *Bulletin of University of Agricultural Sciences and Veterinary Medicine Cluj-Napoca. Horticulture* 74, 83–92.

Ghosh, S. and Dubey, S.K. (2013) Comparative analysis of K-means and fuzzy C-means algorithms. *International Journal of Advanced Computer Science and Applications* 4, 35–39.

Giurca, D. (2008) Semi-subsistence farming – prospects for the small Romanian farmer to choose between a 'way of living' or efficiency. *Agricultural Economics and Rural Development* 5, 215–230.

Greene, W.H. (2000) *Econometric Analysis*, 4th edn. Prentice Hall, New York.

Gu, C. (2013) *Smoothing Spline ANOVA Models*. Springer, New York.

Gutiérrez, J.M. and Hernández, M.A. (1997) New recurrence relations for Chebyshev method. *Applied Mathematics Letters*, 10, 63–65.

Henke, R., Benos, T., de Filippis, F., Giua, M., Pierangeli, F. and Pupo d'Andrea, M.R. (2018) The new Common Agricultural Policy: how do member states respond to flexibility? *Journal of Common Market Studies* 56, 403–419.

Howitt, D. and Cramer, D. (2005) *Introduction to Research Methods in Psychology*. Pearson Education, Harlow, Essex.

Kerenidis, I., Laplante, S., Lerays, V., Roland, J. and Xiao, D. (2012) Lower bounds on information complexity via zero-communication protocols and applications. In: *Proceedings of the 2012 IEEE 53rd Annual Symposium on Foundations of Computer Science*. IEEE Computer Society Washington, DC, pp. 500–509.

Ketchen, D.J. and Shook, C.L. (1996) The application of cluster analysis in strategic management research: an analysis and critique. *Strategic Management Journal* 17, 441–458.

Knight, D.K. (2010) *România şi Politica Agricolă Comună*. EcoRuralis, Cluj-Napoca, Romania.

Lawal, B. (2014) *Applied Statistical Methods in Agriculture, Health and Life Sciences*. Springer, New York.

Lazăr, C. and Lazăr, M. (2016) Trends in the evolution of Romania's agricultural resources in the context of sustainable development. In: Andrei, J.V. (ed.) *Food Science, Production, and Engineering in Contemporary Economies*. IGI Global, Hershey, Pennsylvania, pp. 146–175.

Luca, L. (2009) O ţară şi două agriculturi. România şi reforma Politicii Agricole Comune a UE. Available at: http://www.ecoruralis.ro/storage/files/Documente/CRPE_Policy_Memo_no.4_RO.pdf (accessed 4 February 2017).

Nazzaro, C. and Marotta, G. (2016) The Common Agricultural Policy 2014–2020: scenarios for the European agricultural and rural systems. *Agricultural and Food Economics* 4, 16.

Official Journal of the European Union (2012) Tratatul privind Funcţionarea Uniunii Europene (versiune consolidată). Available at: http://eur-lex.europa.eu/legal-content/RO/TXT/PDF/?uri=CELEX:12012E/TXT&from=RO (accessed 30 September 2017).

Pagh, R., Silvestri, F., Sivertsen, J. and Skala, M. (2015) Approximate furthest neighbor in high dimensions. In: Amato, G., Connor, R., Falchi, F. and Gennaro, C. (eds) *Similarity Search and Applications. Lecture Notes in Computer Science*. Springer, New York, pp. 3–14.

Peyret, R. (2002) *Spectral Methods for Incompressible Viscous Flow*. Springer, New York.

Popa, M. (2008) *Statistică pentru psihologie: teorie şi aplicaţii SPSS*. Polirom, Iaşi, Romania.

Postoiu, C. and Buşega, I. (2015) Romania's agriculture and its role in the convergence process. *Global Economic Observer* 3, 33–43.

Rotariu, T., Culic, I., Bădescu, G., Mezei, E. and Mureşan, C. (2006) *Metode statistice aplicate în ştiinţele sociale*. Polirom, Iaşi, Romania.

Saint-Arnaud, S. and Bernard, P. (2003) Convergence or resilience? A hierarchical cluster analysis of the welfare regimes in advanced countries. *Current Sociology* 51, 499–527.

Sánchez-Garcia, R.J., Fennelly, M., Norris, S., Wright, N., Niblo, G., Brodzki, J. and Bialek, J.W. (2014) Hierarchical spectral clustering of power grids. *IEEE Transactions on Power Systems* 29, 2229–2237.

Skogstad, G. and Verdun, A. (2009) The Common Agricultural Policy: continuity and change. *Journal of European Integration*, 31, 265–269.

Swain, N. (2013) Agriculture 'East of the Elbe' and the Common Agricultural Policy. *Sociologia Ruralis* 53, 369–389.

Swinnen, J.F.M. (2008) *The Perfect Storm: The Political Economy of the Fischler Reforms of the Common Agricultural Policy*. Centre for European Policy Studies, Brussels, Belgium.

Thurston, J. (2008) The CAP and Europe's subsistence farmers. Available at: http://capreform. eu/the-cap-and-europes-subsistence-farmers/ (accessed 21 January 2018).

Tudor, M.M. (2015) Small scale agriculture as a resilient system in rural Romania. *Studies in Agricultural Economics* 117, 27–34.

Walde, J. (2014) Analysis of variance (ANOVA). Available at: https://www.uibk.ac.at/statistics/ personal/janettewalde/lehre/phd_biology/ss2011/analysis_of_variance.pdf (accessed 19 November 2017).

Wharton, C.R. (ed.) (1969) *Subsistence Agriculture and Economic Development*. Routledge, New York.

World Bank (2008) World Development Report 2008. Agriculture for development. Available at: https://siteresources.worldbank.org/INTWDR2008/Resources/WDR_00_book.pdf (accessed 7 November 2018).

World Health Organization (2017) WHO/Europe approaches to obesity. Available at: http:// www.euro.who.int/en/health-topics/noncommunicable-diseases/obesity/obesity (accessed 1 December 2017).

Zaharia, M. and Gogonea, R.-M. (2009) *Econometrie. Elemente fundamentale*. Editura Universitară, Bucharest, Romania.

Zaharia, M. and Oprea, C. (2011) *Elemente de Analiza Datelor şi Modelare Utilizând Excel*. Editura Universitară, Bucharest, Romania.

Zaharia, M. and Popescu, C. (2015) Comparative analysis on the influence of the economic crisis on education in some European Countries. *International Journal of Synergy and Research* 4, 63–83.

Zăpodeanu, D. and Popovici, I.-F. (2006) Instrumente financiare ale Politicii Agricole Comune. *Annales Universitas Apulensis Series Oeconomica* 2, 1–7.

Index

Page numbers followed by 'n' refer to notes.